Community and Capital in Conflict

Community and Capital in Conflict

Plant Closings and Job Loss

Edited by John C. Raines,
Lenora E. Berson and
David McI. Gracie

Temple University Press PHILADELPHIA

Temple University Press, Philadelphia 19122

Published 1982
Printed in the United States of America

Library of Congress Cataloging in Publication Data

Main entry under title:
Community and capital in conflict.
Includes index.
1. Plant shutdowns—United States—Congresses. 2. Unemployment—United States—Congresses. 3. Industries, Location of—United States—Congresses. 4. United States—Economic policy—Congresses. I. Raines, John C. II. Berson, Lenora E.
III. Gracie, David McI., 1932–
HD5708.55.U6C65 338.6′042 82-707
ISBN 0-87722-270-3 AACR2

Absentee control changed the corporation's view of the surrounding community. Communities were viewed as locations for plants which were appropriate only so long as the location provided economic advantages. When such a location was no longer economically rational, the plant was closed and facilities built elsewhere.

This economic calculation creates a fundamental conflict between the calculus of corporate welfare and local community welfare.

Professor Robert N. Stern of Cornell University, testimony before the Committee on Small Business, House of Representatives, February, 1980

The year it shut down the Shoemaker Avenue plant, Federal Mogul reported record sales and more than $14 million in profits. The company's employees, however, didn't fare so well. In the aftermath of the closing, at least seven of Jim Farley's fellow workers took their lives.

Don Stillman, "The Devastating Impact of Plant Relocations" ("Working Papers for a New Society," July–August, 1978).

Members of the Working Group of the Center for Ethics and Social Policy

CAROLYN TEICH ADAMS, Assistant Dean, College of Liberal Arts, and Associate Professor, Urban Studies Department, Temple University.

LENORA E. BERSON, Author, community activist, and lecturer at the Center for Contemporary Studies, Temple University.

DAVID McI. GRACIE, Protestant campus minister at Temple University; Executive Director of the Center for Ethics and Social Policy.

PAMELA HAINES, Staff member, Center for Contemporary Studies, Temple University.

ARTHUR HOCHNER, Assistant Professor, Industrial Relations and Organizational Behavior, Temple University.

GARY KLEIN, Doctoral candidate, Department of Sociology, Temple University.

LYNNE KOTRANSKI, Project Director, Division of Research and Evaluation, Philadelphia Health Management Corporation.

Douglas Porpora, Doctoral candidate, Department of Sociology, Temple University.

John C. Raines, Associate Professor, Department of Religion, Temple University; President of the Center for Ethics and Social Policy.

Edward Schwartz, Director of the Institute for the Study of Civic Values, Philadelphia.

Daniel M. Zibman, M.B.A., School of Business Administration, Temple University.

The Conference Speakers

RANDY BARBER, Co-director of the People's Business Commission, Washington, D.C.

BARRY BLUESTONE, Associate Professor of Economics and Director of the Social Welfare Research Institute at Boston College.

ROBERT LEKACHMAN, Distinguished Professor of Economics, Herbert H. Lehman College, City University of New York.

CHARLES W. RAWLINGS, Officer for Church and Society on the staff of the Episcopal Diocese of Ohio, Cleveland, Ohio.

GREGORY D. SQUIRES, Editor of the Journal of Intergroup Relations, Adjunct Faculty Member in the Sociology Departments at Roosevelt University and University of Illinois, Chicago Circle.

Contents

Part Two
Philadelphia: The Deindustrialization of an American City

Foreword

David McI. Gracie

This book is a project of the Center for Ethics and Social Policy, which came into being in 1979. The center's purpose has been to bring together academics, leaders in the religious community and other concerned people to study important issues of public policy in an ethical framework.

The Center's working group began with about a dozen people: social scientists, theologians, community activists, and at least one poet. It did not take long for a consensus to emerge in this group concerning the set of problems we felt we had to address first. When Arthur Hochner said, "Let's make it the year of the runaway shop!" we knew the consensus had been expressed. Our work together was to center on the flight of jobs and capital from our city of Philadelphia, from the Northeast, and also from the country as a whole.

Here was a major social problem which we believed was not being studied and discussed with sufficient awareness—that it was an ethical concern as well as a question of economics. Plant closings and continued high unemployment affect the quality of our lives and those of our neighbors, undermine the institutions we serve, and place a tremendous strain on the whole social fabric. Perceiving a serious conflict here between capital and community, we decided to begin to

Barbara Baker: Campaign for Human Development, U.S. Catholic Conference—used with permission

Near the plant gates at the Eaton demonstration, May 18, 1981.

analyze and assess that conflict and raise the question of what should be done to serve the public good.

In January, 1981, the Center for Ethics and Social Policy sponsored a conference. Its title was: "Reversing the Flight of Jobs and Capital: Ethics and Economic Decisions." The major presentations at that conference appear in the first section of this volume, with an introductory chapter by Carolyn Teich Adams. Studies which concentrated on the Philadelphia area were commissioned as well. They appear in the second section of the book, edited by Lenora E. Berson and preceded by her history of economic planning in the city of Philadelphia. The concluding section is devoted to discussion of the ethical issues, although they are certainly not absent from any other chapter of this book.

The photographs are provided by the Delaware Valley Coalition for Jobs and the Amalgamated Clothing and Textile Workers Union. They show working people who are beginning to struggle against business and political decisions which have left their futures out of account.

We hope that this book and the work of the Center can make a contribution to increasing understanding and building support for that struggle.

We wish to express our gratitude to the Pennsylvania Humanities Council for the full funding of our conference and the partial funding of the preparation of this book.

Part One

Reversing the Flight of Jobs and Capital

Chapter 1

The Flight of Jobs and Capital: Prospects for Grassroots Action

Carolyn Teich Adams

To anyone acquainted with economic trends in the aging industrial heartland of the United States, Philadelphia presents a depressingly familiar profile. Since 1969, when the Bureau of Labor Statistics first started compiling seasonally-adjusted figures on the region's employment, the city of Philadelphia has lost about 130,000 jobs, or 14 percent of its employment (see Figure 1-1). Naturally, as the absolute size of the job base has shrunk, the size of the losses suffered each year has abated. But the losses have continued. In 1980 the city lost another 3,800 jobs. In fact, the most recent year (1981) was particularly painful for the entire region; for the first time since the recession of 1974–75 the eight-county metropolitan area as a whole lost jobs. Nor do the BLS figures in Figure 1-1 tell the whole story. They show "net" job loss, or the final count when total losses have been partially offset by new jobs created. Thus, the total number of jobs either eliminated altogether or moved out of the city is even larger than the net figures suggest.

Moreover, the impact of job loss has been disproportionately severe in the manufacturing sector, while the creation of new jobs has been disproportionately in the service

Figure 1-1. Total Number of Jobs in Philadelphia and the 8-County Metropolitan Area

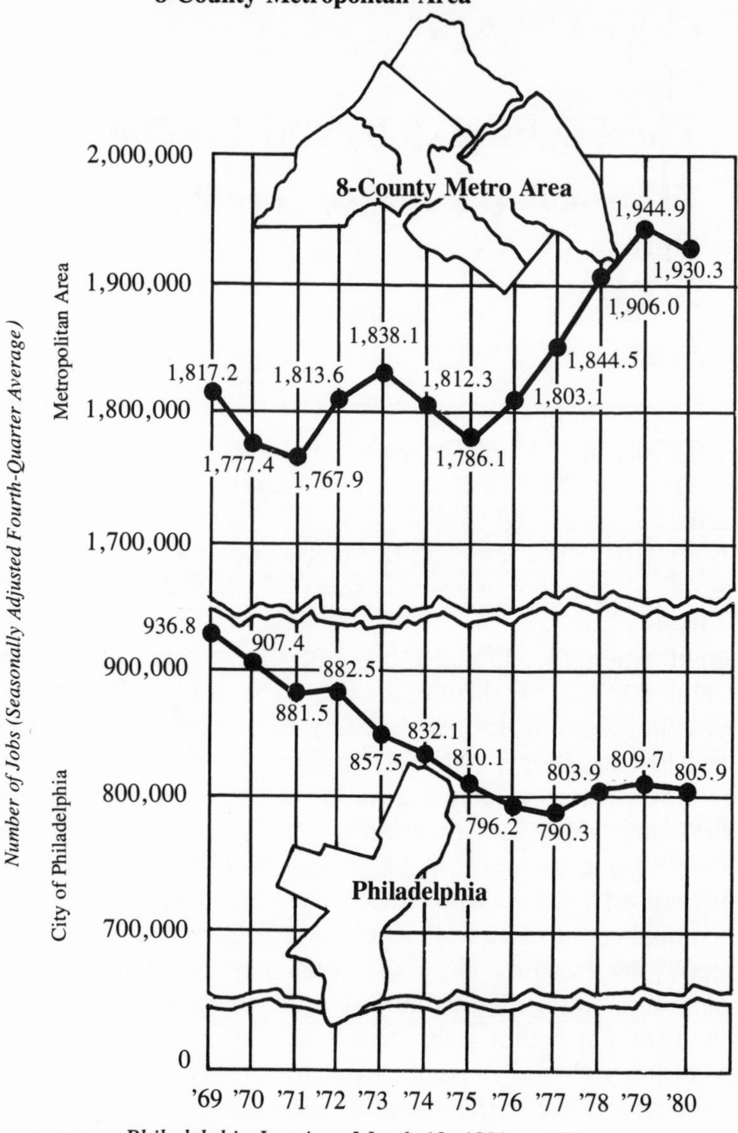

SOURCE: *Philadelphia Inquirer*, March 12, 1981, p. 12A.

sector. Obviously, this shift in the region's employment base away from manufacturing and toward services has imposed disproportionate costs on industrial workers and on the communities where they live.

These debilitating trends—their causes, dimensions, and their meaning for the people living in the affected communities—are the subject of the papers in this volume. The authors were participants in the conference held in Philadelphia in January 1981. The conference topic was not a new one.

For at least fifteen years Philadelphia's Chamber of Commerce, the Penjerdel Corporation, the Greater Philadelphia Partnership, and other business and civic groups have sponsored symposium after symposium on local economic problems. On countless occasions the city's bankers, industrialists, corporate economists, real estate developers, and utility company executives have assembled to deplore the flight of jobs from the city and to pledge their cooperation to stem this tide.

But while the subject of this volume is distressingly familiar, the perspective is new. For these papers emerged from a conference sponsored, not by corporate or civic leaders, but by a small group of church people, neighborhood activists, and academics whose views on, and stake in, the problem of economic decline were different from those of the city's business community. Their emphasis was not so much on the deterioration of corporate earnings or the deterioration in the local tax base; rather it was on the deterioration in the city's neighborhoods, its housing and services, and even the deterioration in family life that can be traced to the loss of jobs. Many of the conference participants were people whose lives have been directly affected by the flight of investments and jobs from Philadelphia.[1] Thus the proceedings had an urgency that is missing when government and business leaders gather to discuss this same set of issues.

Yet this urgency did not blind the organizers and participants in the conference to the undeniable fact that many of the forces responsible for the trend are beyond the reach of local citizens' groups. Not only are decisions to move jobs and capital made by national and international firms that are bound by no special obligations to any particular location, but corporate decision makers, as several conference speakers emphasized, are themselves also caught up in economic and political structures which dictate many of their decisions.

The realization of the tremendous odds working against any grassroots effort lent a kind of "heroic" quality to much of the discussion. That quality is perhaps best illustrated by the comments of Rev. Charles Rawlings, a veteran of three years experience in organizing opponents of plant closings in Youngstown, Ohio. Even as he urged his audience to build upon the Youngstown experience, Rev. Rawlings admitted wearily to

> a sense that events are rapidly moving out of our grasp. The drift of things, politically and economically, is so severe that there is now a question as to whether or not we can reverse it, get hold of it, and change it. . . . the day is far spent.

Rawlings cautioned that in Youngstown even concerted action by a large religious, civic, and professional coalition did not prevent 12,000 workers from losing their jobs.

Rawlings' admonition against undue optimism raised the fundamental question about this conference, or any similar effort at grassroots organizing: apart from the urgency of the matter for their communities and families, do local citizens have any realistic hope that their actions will make a difference? Or is this problem to be either perpetuated or resolved by forces that are beyond the reach of local citizen activists, no matter how well organized?

Goals of the Conference

Putting the Issue on the Public Agenda

The fundamental purpose of the conference was simply to raise as a public policy issue the flight of capital from the region; that is, to get the issue placed on the public agenda. Conference organizers were particularly interested in raising the question of whether local, state, and national governments can take action to restrain the mobility of capital.

Simply to raise this question for public debate may at first glance seem easy to do. It is not. Dozens of previous conferences had been organized by Philadelphia's business and government leaders, all treating the general problem of the city's economic decline. Yet none of them had focused on the prospects for restraining corporations from moving their plants and jobs away from the region. Rather, at such conferences the movement of capital had been taken as a given—an inevitable, even natural feature of the economic environment. The issue had, therefore, been not how to prevent it, but either how to lure investments from someplace else or how to lessen the disruptive effects of the losses that the city must inevitably sustain.

In describing his efforts to mobilize church members in Youngstown against plant closings, Rev. Rawlings reported that he found it surprisingly difficult to confront the silence of church congregations about what was happening in their own communities. He attributed this silence to a lack of political consciousness on the part of church people. Another way of explaining the silence is to be found in the concept of "mobilization of bias," enunciated by E. E. Schattschneider several decades ago. Succinctly put, Schattschneider's view is that any community's politics display a bias toward airing some kinds of policy conflicts but not others: "Some issues are organized into politics while others are organized out."[2]

In their critique of community power studies, Bachrach and Baratz elaborated upon this idea further, arguing that one of the most important forms of power that a local elite can exercise is the ability to limit public debate to "safe" issues which do not fundamentally challenge their interests. By "manipulating the dominant community values, myths, and political institutions and procedures," elites keep the truly threatening issues off the public agenda altogether and thereby avoid having to exercise their power openly.[3]

The question of whether governments should take steps to discourage employers from moving their investments out of the city is one which both government and business leaders would prefer to keep off the public agenda. Why? The answer involves that constellation of public and governmental attitudes toward business which is usually labeled the "business climate." In a survey reported in 1979 by the Joint Economic Committee of Congress, a national sample of businesses was asked what characteristics of communities were most important in influencing their location choices. More respondents picked "city government attitudes toward business" than any other factor. (Factors ranking below this top choice included "crime level," "adequacy of public facilities," "market demand for product or service," and even "cost of energy.")[4] In 1977 *Fortune Magazine* emphasized the importance of business climate in the Sunbelt's boom:

> It's booming in great part because it's pro-business—and Northern cities by and large aren't. Much of the region is a repository of traditional American values—patriotism, self reliance, re-

> spect for authority—and both racial disorders and street crime are relatively rare. . . . It is also true that many Sunbelt states solicit new business with zeal and skill . . . the most effective form of aggressive behavior . . . is the joining together of politicians and businessmen to shape state laws that favor business.[5]

As this observation from *Fortune Magazine* makes clear, a region's business climate encompasses not only governmental attitudes and actions, but also the general political culture of the region. Mollenkopf has suggested, for example, that it is not simply the pro-business climate of the Sunbelt that has lured businesses away from the Northeast. He believes that often businesses move in order to escape from the high levels of militant community activism in Frostbelt cities.[6]

If the perception of the local business and political climate is so important in determining choices of location, then one might argue that the very mention in a public forum of possible governmental restraints on a company's right to shift its investment could make the city less attractive to investors. It is hardly surprising then that conference organizers had great difficulty securing from both local government and the local businesses the participation of representatives who were not sure the subject should be discussed at all.

Challenging Myths

In addition to this primary goal of raising the issue publicly, a second aim of the conference was to challenge a series of commonly-held myths about the flight of jobs and capital from the Northeast. Among them are the following.

Myth #1. That small independent firms are the most vulnerable to the rising costs of doing business in the Northeast, and therefore are the most likely to close their plants.

One common assumption about the loss of jobs in the Northeast is that it has resulted primarily from the failures of small, independently-owned firms. The high failure rate of small businesses is a well-known statistic, and it seems logical to presume that firms without a parent company to support them in difficult times will more likely go under when faced with increasing pressure from militant unions, skyrocketing energy costs, etc.

Bluestone's contribution challenged that simple logical progression. He reported that his own work examining plant closings in Massachusetts and New England from 1969 to 1976 does indeed show a significant failure rate among independently-owned firms. But when he compared their failure rate to the rate at which large conglomerates closed down their facilities in the same region, he found that they were closing down plants at a higher rate than either the small independents or the locally-based corporations. More importantly, the closings of plants by conglomerates, he found, had a more adverse impact on regional employment because conglomerates were less likely than other kinds of owners to create new jobs to balance the closings. Over the period 1969 to 1976, for every ten jobs lost in New England through the closing of an independently-owned business, six new jobs were created by small independents. During that same period, for every ten jobs phased out by conglomerates, only four new ones were created by conglomerates.[7]

Bluestone's report on his work in New England and Massachusetts is confirmed by Hochner's analysis of Philadelphia plant closings presented in this volume.[8] Here Hochner found that conglomerate owners have removed many more jobs than either small independents or locally-based corporations. Hochner concludes that the larger the firm, the more mobile the capital. The message to this audience was clear: the largest part of the problem of capital and job mobility results from decisions made in the headquarters of national and international conglomerates. Local efforts to assist small businesses, however popular and well-organized at the local level, are not likely to change the decision-making patterns of the large absentee owners.

Myth #2. That plants close because they are no longer capable of generating a profit for their owners.

It is a truism that the costs of doing business are rising in the northeastern United States. There can be no doubt that wages, energy costs, land costs, and taxes have steadily increased in the older industrial cities and that these costs, taken together, have had an adverse impact on the profits of enterprises located in this region. But is it true that these rising costs have virtually eliminated the profits of the thousands of plants which have closed in the region over the past ten years? Is it true, in other words, that plants are shut down because they are no longer capable of making a profit?

Again, Bluestone called into question this widely shared assumption. His investigations and those of William F. Whyte at Cornell, indicate that, in fact, profit-making plants are frequently shut down.[9] Large corporations, especially conglomerates, often close plants not because they are unprofitable but because the rate of return on the investment is not as high as could be gained from other forms of investment. This fact of economic life—that capital seeks the highest available rate of return—is not surprising nor is it particularly sinister. What is disturbing is that by their own management decisions corporations sometimes contribute to a plant's declining profits, which then become the justification for a shutdown.

Here again, large corporations and conglomerates provide

the most frequent and the most dramatic illustrations. Bluestone cited numerous examples of parent companies appropriating profits made by their subsidiaries, only part of which they reinvest in the facility which produced that profit. This practice, Bluestone related, is common enough to have acquired a shorthand label in business circles—it is known as milking "cash cows." One negative result of this practice is that subsidiaries have little opportunity to build up reserves for use in years when profits dip. In such years, they may be forced to borrow money in local capital markets, paying commercial lending rates if they can secure the loans at all. An even more damaging possibility is that parent companies may take the profits from a subsidiary, not even reinvesting the minimum necessary to maintain the subsidiary at a profitable level. By deferring purchase of modern equipment, the parent company virtually insures that the subsidiary's profits decline. This form of redistribution of capital between plants—even between unrelated industries—is possible only for large corporations and is particularly common in widely-diversified conglomerates. The same is true of other kinds of destructive management practices cited by Bluestone, e.g., the practice by some parent companies of forcing subsidiaries to buy raw materials from distant suppliers owned by the conglomerate, instead of encouraging them to buy raw materials at the cheapest price.

In short, Bluestone observed, it is much too simple to see plant closings as the inevitable result of rising costs that wipe out profits. The conscious policies of managers may cause even profitable enterprises to close because they are not "profitable enough," or their policies may deprive once-profitable plants of the continuing investments they need in order to stay "profitable enough."

Myth #3. That the economic advantages enjoyed by the Sunbelt in recent decades result simply from the superior market conditions prevailing in the Sunbelt, i.e., cheaper labor, energy, land, etc.

It is a commonly-held view, even in the large cities of the older Northeast, that the Frostbelt is losing out to the Sunbelt because objective market conditions have dictated such a shift. The perception is that the Sunbelt's advantage in terms of the lower cost of doing business there has simply forced investors to chose Sunbelt locations.

A somewhat different picture, however, was painted by a Congressman who addressed the conference on the subject of regional decline. Representative Robert Edgar, a Democrat representing a suburban Philadelphia county, is currently chairman of the Northeast-Midwest Congressional Coalition. This bipartisan caucus, including members of Congress from 17 states in the Northeast and Midwest, was formed in 1976 when researchers at the *National Journal* began to publish figures showing that the Frostbelt was being seriously short-changed by federal spending patterns. In fiscal 1975, the *National Journal* calculated, the fourteen states of the Northeast and Great Lakes region had sent $29.4 billion more to Washington in federal taxes than they had received in federal expenditures. In that same year, the South and West had gained $22.2 billion in federal expenditures above what they had contributed in federal taxes.[10] This revelation led to further analysis of the discrepancies in federal spending in various sectors (defense installations, federal civilian employment, highway construction, transfer payments of various kinds, etc.). The conclusion of many observers was that far from suffering a spontaneous decline, the Frostbelt was having its resources systematically siphoned away to build up the Sunbelt.[11] The Northeast-Midwest Congressional Coalition has worked during the last five years to try to reverse this pattern of federal expenditures.

In addition, the Coalition has attempted to change federal policies intended to influence the pattern of private business investments. Congressman Edgar gave particular emphasis in his remarks to two features of the federal tax code that have encouraged the shift of private investment capital away from the Northeast-Midwest region: accelerated depreciation allowances and the investment tax credit. Figures collected by the Northeast-Midwest Institute (the Congressional Coalition's research arm) show that from 1970 to 1977 investment in manufacturing equipment and capital plant in the Frostbelt increased by only 23 percent, compared to an increase of 74 percent in the South and West. Even more startling is the finding that in *urban* areas of the Northeast and Midwest, such investments actually declined by 2 percent.[12]

While it did not create this trend, the federal tax code has reinforced it. First, the tax code allows businesses to calculate their tax liability using accelerated depreciation allowances on new structures that they purchase. This effectively accelerates the deductions that businesses are allowed to claim to reduce their taxable income in the first few years after they make the investment. In contrast, businesses must depreciate used plants, machinery, and equipment more slowly. The result of this discriminatory treatment is to encourage businesses to build new plants rather than buying existing plants.[13] Similar encouragement is offered by the government's system of tax credits to businesses. Up to 1978 federal tax laws allowed businesses to claim a tax credit of 10 percent of their expenditures for buying new machinery and equipment, while allowing no credit for rehabilitating the existing capital plant. While the Revenue Act of 1978 introduced tax credits for rehabilitating existing machinery and equipment, it limited the credit on used equipment to $100,000 in a single tax year. No such limitation is placed on credits for new machinery.[14] Once again, the effect is to lower the cost of moving out of existing plants and buying entirely new plants, thus making it easier for firms to shut down plants in the older parts of the country and reopen in the South and West.

Representative Edgar's remarks dispelled the myth that the Sunbelt's economic advances have been won in pure and open competition with the aging industrial heartland. On the contrary, the South and West have benefited from significant federal subsidies.

Myth #4. That local tax abatements are a critical inducement in attracting and holding businesses

Perhaps because of the ubiquity of tax concessions offered by cities and states to lure businesses into their jurisdictions, most of us assume that local and state taxes must be an important consideration influencing business location decisions. Yet several speakers during the conference emphasized that virtually all systematic investigations of the question show otherwise. In fact, the research on the question is almost unanimous in proclaiming that the local tax abatements and exemptions that are part of the traditional package used to lure newcomers, are actually unimportant considerations in the calculation made by corporate decision makers.[15] Moreover, all local and state taxes paid by industry are deductible from the firm's income for federal tax purposes, so that a reduction in a firm's local taxes will increase its federal tax liability.

Moreover, it is questionable whether tax abatements offered to businesses expanding their current operations actually generate more jobs. A study of Philadelphia businesses issued in December 1980 concluded that tax abatements to businesses for expansion are a costly and unreliable way to combat hard-core unemployment. Offering tax forgiveness to an expanding firm does not guarantee that the company will hire more workers. In fact, in some cases, expansion may be accompanied by increased mechanization, thereby further reducing the number of workers employed by the firm. In such cases, public subsidies are being used to help employers eliminate jobs.[16]

If local tax abatements are an unreliable way to generate jobs within expanding firms and an insignificant factor in corporate decisions to relocate their plants, why are they so common? One explanation was provided by columnist Neal Pierce, after he interviewed General Motors' spokesman on plant locations about the new plant that General Motors is building in Detroit. In a celebrated case of municipal boosterism, Detroit offered GM a newly-assembled 468-acre site from which 1,360 homes and 143 businesses were to be cleared at a total cost to the government of $199 million. But beyond this subsidy, GM demanded and received from Detroit a twelve-year tax abatement that would cost the city over $100 million in tax revenues. After acknowledging that the tax abatement represented an insignificant factor in choosing a plant location, the GM spokesman explained that his company had nevertheless demanded such a concession from Detroit simply because it was allowed under Michigan law: "We'd be unresponsive to our shareholders not to request it."[17] The Detroit case suggests that while tax concessions are a relatively unimportant element in the corporation's financial calculations, a city's unwillingness to grant them may be construed by corporate decision makers as a sign of an unfavorable "business climate."

Myth #5. That the nation achieves greater efficiency by allowing, even encouraging capital mobility away from the aging industrial cities

Only two weeks before the conference was convened, the outgoing Carter administration in Washington had received the final report of its Commission for a National Agenda for the 80's. Appointed by Carter in 1979, the McGill Commission (named for its chairman, William J. McGill, former president of Columbia University) had surveyed a wide range of economic and social problems and proposed federal policy directions for the decade of the 1980s. The commission's urban panel had called for the redirection of federal urban policies away from "place-oriented" aid to "people-oriented" projects. The commission's urban panel argued that "aiding places and local governments directly for the purpose of aiding people indirectly is a policy emphasis that should be re-examined at a time when successes are so few and public resources so meager." According to the report, "cities are not permanent," and policies which treat them as permanent are doomed to fail because they run against the tide of market forces. Instead, the federal government should "let the market function and then assist people to adjust." The major policy recommendation of the report was, therefore, to encourage the migration of people from the Frostbelt, where jobs are disappearing, to the Sunbelt, and to offer re-training to the migrants to improve their chances of finding employment in their new homes.[18]

The assumption underlying this recommendation was that investing money in the Frostbelt is counterproductive because economic efficiency dictates the movement of capital out of the Frostbelt. After all, haven't businesses demonstrated that this movement maximizes their profits? The difficulty with this reasoning is that it equates the most efficient use of public capital with the most efficient use of private capital. It ignores the fact that public money is often used to pay the costs imposed on communities by businesses.

The large-scale migration of plants and workers to new areas of the country causes the virtual abandonment of many industrial facilities in the Northeast and Midwest. Not only are the plants themselves duplicated in the new locations, but the infrastructure needed to serve these plants must also be recreated at the new site. New investments must be made in roads and other transportation systems, as well as new schools, hospitals, utilities, waste treatment facilities, and other community facilities to serve both the migrating workers and the migrating industries. On the other hand, the communities left behind also need to make new investments, to adjust the physical infrastructure and land-use patterns to suit new industries and new technologies if, indeed, they are able to attract new industries to replace those leaving. Moreover, these communities must either invest in retraining the workers left behind, or else bear the tremendous costs of sustaining large numbers of unemployed. Nor can one argue that from a national point of view, the cost of coping with unemployment generated in the declining regions is balanced by a drop in unemployment in the high growth cities of the Sunbelt. The work of a number of economists has undermined the assumption that rapidly-growing areas have lower unemployment rates than other urban areas. "In fact, the tendency is for rapid growth to be associated with higher rates of unemployment."[19]

None of these costs is borne directly by the businesses whose decisions impose the costs on communities. Nor are all these costs borne by a single unit of government; rather, they are absorbed by a combination of several local jurisdictions, state governments, county governments, and the federal government. Hence, the full extent of these costs is rarely tallied. From a national perspective, however, they are substantial. When taken together, they call into question the superficial assumption by the McGill Commission and others that we make the most efficient use of our national resources by encouraging capital mobility.

Emphasizing the Ethical Dimension of the Problem

Yet another goal of the conference organizers was to stress the ethical dimension of both public and private decisions regarding capital shifts. This is a dimension that is often missing from discussions of local economic development, especially those guided by economists. Most economic analyses of local and regional economies rely on aggregate statistics and equilibrium models. They view the local economy as a system whose stability depends upon maintaining a certain balance between numbers and kinds of jobs and residents lost, versus numbers and kinds of jobs and residents gained. Because it looks at overall figures instead of individual families and communities, systems analysis frequently has an antiseptic, even mechanical quality.

The focus of this particular conference was on value questions rarely raised in these kinds of analyses. One theme running through the day's discussion was the question of whether our society has a moral obligation to minimize the waste of resources involved in capital shifts. This question is related to the point elaborated above; namely, that the movement of plants away from the Northeast and Midwest entails the abandonment of already-existing infrastructure and its duplication elsewhere. At a time when we are increasingly conscious that our material resources are finite, can we afford to use them up in this way?

An equally important question raised during the day was whether capital mobility in our society places disproportionate burdens on some groups. This issue was addressed at the conference most directly by a researcher for the U.S. Commission on Civil Rights in Chicago, Gregory Squires. He reported on a Civil Rights Commission study of all plant relocations from central cities to suburbs in Illinois from 1975 to 1978. Taken together, these relocations resulted in unemployment for 25 percent of the plants' minority workers, but for only 10 percent of their white workers. The Civil Rights Commission found that minority employees were less likely than white employees to know about the relocation in advance and less likely to be offered aid for relocation. Because minority workers were concentrated in non-managerial positions, they were less likely to get any of the benefits extended to managerial employees, to cushion the damaging effects of the relocation. Such findings as these, Squires related, have led some civil rights attorneys to file lawsuits against runaway plants, charging that a relocation which has a disproportionate impact on minority employees violates Title VII of the 1964 Civil Rights Act.[20]

A related question on which we have little data concerns the disproportionate impact of plant relocations on women workers. Typically, women workers must work closer to their homes than men because of their role as the primary caretakers of home and children.[21] Women are less likely than men to be able to follow their jobs when plants relocate. Typically, they earn less than their husbands; and it is, therefore, the husband's job which takes priority in the family's decision about where to live. Because they have less flexibility than men in choosing a workplace, it seems reasonable to assume that women workers are more adversely affected by plant relocations.

A major concern expressed by many participants involved the issue of corporate responsibility. Does an employer have an obligation to notify both employees and the wider community well in advance about plans to shut down a plant? The argument that both employees and communities have an inherent right to know about such plans in advance is in some ways parallel to the reasoning that has led to other recent kinds of legislation involving corporate responsibility to the community. For example, the federal Home Mortgage Disclosure Act of 1975 was premised on the community's right to know where local banks were lending their mortgage money; this law required that banks disclose to the public the number and amount of residential mortgages and home improvement loans made to borrowers in different neighborhoods of the cities where they have offices. The passage of this law was only the first step in communities' efforts to stop redlining, or the practice by lenders of refusing to grant mortgages for properties located in certain "undesirable" neighborhoods. The law does not by itself prohibit redlining, but it does at least guarantee the public's right to know about the pattern of private lending in the city, so that people whose neighborhoods are adversely affected may take some action.

A second example of such right-to-know legislation is Philadelphia's ordinance passed by the city council in early 1981 forcing local industries to disclose what toxic chemicals their plants store, use, or emit into the air and water. In addition to promulgating regulations on industrial emissions and storage of 450 toxic substances, the ordinance clearly established that firms have an obligation to supply private residents of the surrounding community with the information they need to cope with any adverse impacts on their homes and families.

Both of the above examples illustrate the principle, now rather widely accepted, that private companies do have some obligation to disclose to the communities in which they do business information about actions that may damage those communities. Following this same line of reasoning, could we not conclude that companies have an obligation to disclose in advance their plans to shut down plants which employ a significant number of workers and whose closing will damage the local economy?

Exploring Alternative Responses to the Problem of Capital Flight

The approaches to the problem of runaway shops advocated by the various speakers at the conference ranged from local, even neighborhood-centered activities, to national responses. The actions suggested by the various speakers aimed at different targets; but each time a target was identified and discussed, it was found embedded in some larger context, and participants were forced constantly to expand the frame of reference within which they understood the problem of capital mobility.

Rev. Charles Rawlings of Youngstown, Ohio, was an organizer who had spent several years trying to arouse the conscience of church-goers in his community to recognize that the loss of jobs by 12,000 factory workers was a loss even for citizens not left jobless by the shutdowns. His approach was to call for more education, even consciousness-raising, so that individuals whose jobs are secure can begin to recognize the connections between their own welfare and that of their unemployed fellow citizens. Without such consciousness-raising, he feared, citizens not directly affected cannot be counted on to care about the issue of runaway shops.

Other participants were less interested in reaching the public at large than in directly influencing the employers. For example, a young organizer representing the Delaware Valley Coalition for Jobs reported on his organization's campaign to persuade Container Corporation and its parent company, Mobil Corporation, not to close a Philadelphia container factory employing over 100 workers. Working alongside the union, the Coalition for Jobs appealed to the company's chief executive officer. While they did not succeed in keeping the plant open, they believe their efforts persuaded the company to increase the severance payment to workers.[22]

Rather than reacting to shutdowns as they are announced, plant-by-plant, some participants wanted to aim their efforts at local government policies that might encourage companies to maintain jobs in the region. Elected officials, because they depend on public support to stay in office, were seen as more logical targets for pressure than corporate decision makers. Persuading city officials to take an active role in sustaining old industries, promoting new industries, and even regulating local businesses in the community's interest, some participants saw as the best attack on the problem.

Still others argued that concentrating on local efforts at "boosterism" and/or local regulation of business is undesirable because it pits communities against one another as rivals for business investment. In such competition for tax ratables and jobs, cities can spend enormous sums to lure new investors or just to retain existing investments. Often the cities forced to offer the highest inducements are those which can least afford them. At the same time, city governments that contemplate any form of regulation—e.g., requirements that plants provide advance notification of any cutbacks or closings, requirements for severance pay to the workers, or "exit fees" to the city government—risk being seen to have a poor business climate. Seeing local inducements as too expensive and local regulation as a suicidal strategy, many at the conference advocated instead state-level programs of industrial development coupled with state regulation of runaway shops.

A case in point cited several times during the day was Detroit's decision, announced in January 1981, to require a machine tool company to pay an exit fee of $360,000 to close its Detroit plant and move to a site in nearby Macomb County. Since Macomb County had offered the company a tax abatement totaling $3.6 million, the company was in effect forced to contribute a tenth of its abatement to the city as compensation for the loss of jobs and tax revenues.[23] Detroit's compensation was made possible by a state law in Michigan that requires a firm to obtain permission to relocate from the community that is losing the firm before it can accept a tax abatement from another Michigan community. The state law is intended to discourage Michigan communities from raiding one another's tax bases.

The emphasis on state legislation, however, was itself questioned by others at the state conference, who argued that the regional rivalry between the states of the Sunbelt and Frostbelt creates a need for national approaches. Among the proponents of national strategies was Randy Barber, who focused his attention on the largely unrecognized source of investment capital inherent in employee pension funds. Owning about 25 percent of the stock in American companies, pension funds constitute the largest single source of investment capital in the economy. As such, they have contributed significantly to the growth of the Sunbelt. But, as Barber reported, these funds are generated disproportionately in the Frostbelt states. Ironically, Northern workers have had their retirement funds used to build the very plants in the South and Southwest which compete so successfully against them, and in some cases even cost them their jobs. The solution, Barber observed, is for workers to exercise more direct control over the investments made with their pension funds. In this way they can tap this rich source of capital to help rebuild the aging industrial base of the Northeast and Midwest.[24]

Another spokesman favoring national initiatives was Congressman Robert Edgar. After highlighting the imbalance in federal spending that has favored the Sunbelt over the Frostbelt for decades, Representative Edgar concluded that this historic imbalance can be redressed only by federal policies which favor the older industrial regions. His policy agenda included:

- targeted tax reductions, as opposed to across-the-board cuts. Of particular importance is a tax reform to give firms exactly the same credit for rehabilitating old facilities in central cities as for building a new plant in a new location.
- a federal policy that would force the oil-rich states of the Sunbelt to share the revenues they gain by taxing each barrel of oil extracted within their borders. Edgar argued that in levying this severance tax on oil, the states of Texas, Louisiana, and Oklahoma are gaining windfall revenues from what is fundamentally a national resource.
- targeted federal procurement policies, so that suppliers based in the Northeast and Midwest are given preference.

The most comprehensive statement of a national strategy was provided by Robert Lekachman. Cautioning his audience that "fighting local causes should not lead us to neglect the larger vision," he argued for nothing less than a national shift to democratic socialism. The democratic socialist alternative that he outlined included a full-employment policy, national economic planning, nationalization of some industries that have performed poorly in the private sector (e.g., health care, housing, energy), and a national incomes policy. Without such large-scale government intervention in the economy, Lekachman warned, we will make no significant progress in solving the problems of unemployment, inequality, and regional imbalances.

All told, the list of possible targets-of-action ranged from individual company executives, through local and state officials, all the way to national policy makers. Each new discussion brought a recognition that the decision makers in question are merely actors in a larger system which to a great extent dictates their behavior. Even at the national level, one must acknowledge that U.S. economic policy is embedded in a world context that constrains our choices.

Obstacles to Grassroots Organizing

While the conference succeeded in raising important public questions and in presenting alternative responses to the problem of capital flight, it also demonstrated the tremendous difficulties in mobilizing for local political action on these issues. There is no doubt that the day's program helped to give participants a greater appreciation of the complexities involved in explaining the loss of jobs and investments from their region. Contrary to a popular belief, the Northeast is not losing jobs simply because we have made it impossible for businesses to make a profit here. The explanation for the drain is much more complicated than that. Even profitable enterprises have been shut down, and the level of taxation has been found in most research not to be a critical factor in the locating decisions made by businesses. These discoveries might be heartening to Philadelphians, but for the realization that must quickly follow. If plants are not being forced out of the region, but are moving instead because of complicated corporate strategies based on other considerations, then it is unlikely that local initiatives will keep them here. Although we can easily imagine ways of organizing at the grassroots level to influence local politicians or even local union leaders, it is more difficult to see how Philadelphians can organize to affect business conditions in the Sunbelt or corporate strategies pursued by national and international conglomerates.

Moreover, the discussion of alternative responses to the problem highlighted the interdependence among all the people involved. Each time a particular set of decision makers was identified as a target for action (e.g., city council members, local firm managers, state legislators), they were seen to be simply operating as part of a larger structure. Ultimately, all are embedded in an international economic order whose shape is beyond the control of national policy makers, let alone local citizen groups. For local groups to pick the target that is most accessible (i.e., local leaders) is to invest their energy with little hope of altering the wider set of institutional relationships that determine local decisions.

Nor is it easy to assign blame, as Bluestone pointed out. The "right" and "wrong" of current economic problems are more difficult to define than they were for the civil rights and anti-war movements of the 1960's. The issues of import restrictions, the Chrysler bail-out, and mechanization of manufacturing by robots and other means, are not easily represented by slogans that fit on picket signs. Hence, the risk taken by conference organizers was that in helping participants to appreciate the full complexity of the problem, they might paralyze rather than mobilize their audience. Lacking easy formulas or even a clearly-defined target, citizens are likely to give up and go home.

One way to move citizens to action despite such complexities is to dramatize the human costs of capital flight. The organizers of this conference struck upon the device of dramatic readings for which the script was reconstructed from actual interviews with four Philadelphia workers, two women and two men. Their accounts of their own struggles to get an education and training, to find their first job, to stay employed, and to make ends meet vividly illustrated the centrality of work in their lives. As Raines' chapter in this volume points out, the cost of unemployment to them and to others like them must be measured not only in the loss of income, but also in family disintegration, alcoholism, and mental illness. The audience's enthusiastic response to the dramatic readings made clear that this portrayal of the problem in human terms gave it an immediacy that was missing from the more abstract discussion of economic policy issues.

Besides demonstrating the complexity of the issue, the conference also made obvious the obstacles to forming a broad-based coalition even at the local level. In particular, the conference raised the question of whether local business interests can be brought into collective efforts to stem the flow of industrial capital from the region. At first glance this may seem a curious concern. After all, local businesses have a strong stake in maintaining the vitality of the local economy. And in fact Berson's chapter in this collection demonstrates that Philadelphia's business community has since 1945 actively promoted urban redevelopment schemes whose major purpose was to maintain the city's business base. Why then would we question business cooperation in the campaign to maintain the city's industries?

The answer is that broad-based coalitions, while they may be easily formed around a very general set of goals, are extremely difficult to hold together once discussion focuses on the details of their program. Nowhere was the potential for conflict more evident than in the panel session which brought together an economic development official from the city government, a citizen activist from the Delaware Valley Coalition for Jobs, and the head of the Greater Philadelphia Partnership, an organization of business and civic leaders with a primary interest in the city's economic welfare. The tensions between these panelists evolved from two fundamental disagreements.

The first was over the question of whether economic development efforts should concentrate on retaining existing jobs or creating new jobs. According to the spokesman from the Delaware Valley Coalition for Jobs, the community's priority must be to retain the jobs that are now located in the region, especially the manufacturing jobs. He compared the average weekly wage in the city's manufacturing industries ($265) with the average weekly wage of retail clerks ($124) or bank tellers ($174), and observed that the large gap between wages in the manufacturing and service sectors leaves us little choice. We must resist the elimination of manufacturing jobs and their replacement with service jobs.

The spokesman for the Greater Philadelphia Partnership vigorously disputed the wisdom of a campaign to retain the existing manufacturing base. Such a strategy, he proclaimed, "flies in the face of global economic reality." That reality, he pointed out, has two undeniable characteristics. First, a shift from manufacturing to services has been occurring since the 1950's in the economy as a whole, but particularly in central cities. Second, investment capital has become more mobile, both nationally and internationally. Firms are closing down all the time, as the result of rapidly-changing technologies and world competition. Our difficulty in the northeastern U.S. is not that we have businesses closing, but that we have fewer new businesses opening than does the Sunbelt. To balance every 100 jobs lost in the Sunbelt through shutdowns or cutbacks, 111 new jobs are created. In contrast, for every 100 jobs lost in the Northeast, only 85 new jobs are created.[25] Thus, the city's economic development officials must formulate a development plan that capitalizes on the regional shift toward a post-industrial economic base. Our task, according to the business community, is to find ways to spawn new jobs, not to try to keep alive our dying industries. Capitalizing on existing trends means appealing to service industries for whom a central city location is an advantage—banks, financial firms, insurance and real estate companies, marketing, advertising, and data processing firms, and similar services. Whether such businesses are likely to provide jobs for the residents of the city who are now unemployed is an open question.

The second basic disagreement emerging from the panel in Philadelphia concerned the best instruments at our disposal to influence business decisions regarding the location and expansion of enterprises in the region. As might be expected, the dispute centered on the question of "carrots" versus "sticks." Again, the lines were sharply drawn between the panelists representing the citizen coalition and the business community, the former favoring state and local regulations regarding plant relocations, while the latter preferred to offer subsidies to lure businesses to the region. Not surprisingly, the spokesman for business applauded an idea that has recently gained adherents in both business and government circles—the free enterprise zone. This approach to reviving inner city economies, which has attracted the support of both liberals and conservatives, provides a combination of incentives to investors to locate new firms in designated sections of the city and to hire local workers. The most elaborate of the proposals would create islands in which new businesses would receive not only generous tax breaks but also shelter from a series of government regulations, ranging from health and safety and pollution regulations to the minimum wage. The idea is described by its proponents as "greenlining" inner cities, as contrasted with "redlining" them. It will lure private entrepreneurs to invest in an environment where many of the normal constraints and costs of doing business have been lifted.

When the idea of the free enterprise zone was introduced in the U.S. Congress in 1980, it attracted support from such varied political groups as the NAACP, the Urban Coalition, and the conservative Heritage Foundation. In Philadelphia, as in many cities around the country, local business organizations like the Greater Philadelphia Partnership have picked up the idea and are advocating local versions of the enterprise zone. Even before federal tax concessions become available under any federal program, the Partnership would like to see the burden of state and local taxes and regulations lightened for firms willing to relocate and hire workers in Philadelphia's inner city. The main argument for the enterprise zone is the apparent inability of government programs to revive the economies of central cities. Given government's impotence, why not give the private sector an opportunity?

The business community's enthusiasm for the enterprise zone concept is matched by an equally strong aversion to any legislation, by either state or local governments, that would restrain entrepreneurs from closing their plants. Even the hint of such "punitive" legislation, they claim, will detract

Jack Franklin: Delaware Valley Coalition for Jobs

Three hundred people join a "March to Save Our Jobs and Neighborhoods" in Philadelphia, March 15, 1980.

from a city's reputation as a good place to do business. If plants cannot close when they want to close, they will never open in the first place.

On which side of these disagreements did the city official come down? To a very great extent, he supported the position of the business community. Solving the city's economic problems requires far more resources than the city government itself commands. Therefore, virtually all of the city's development schemes must begin by finding a private partner to contribute private sector resources. This emphasis on partnership with the private sector, long a feature of the city's development programs, means that most such programs rely on carrots rather than sticks. That is, they offer subsidies, concessions, and other incentives rather than regulatory actions.

Better than any other part of the day's program, this debate over Philadelphia's development strategy illustrated the uphill fight that faces organizations trying to slow the flight of capital from one region of the country to another, and from one sector of the economy to another. Unions and their supporters have a very serious symbolic disadvantage because their position in this debate appears to be both conservative and protectionist. Ours is a political culture that has associated opportunity with mobility, from the time of the settlers pushing the frontier westward, through the period of the great migrations that brought hundreds of thousands of Europeans to American cities to seek their fortunes. Recently, the massive migration from America's inner cities to the suburbs has reflected once again that to many Americans "moving up" implies "moving out."

In contrast with this tradition that ties opportunity to mobility, the unemployed workers living in declining communities of the Northeast and Midwest have begun to see things differently. They have begun to question whether economic progress necessarily depends on high levels of mobility of both capital and workers. By calling for measures to restrain the flow of capital out of their communities, they are in effect trying to preserve the *status quo*, to slow the pace of change. Their position is thus seen by opponents as a conservative, even protectionist position, while the position of the business community is portrayed as the more dynamic and progressive viewpoint.

How should unions, neighborhood organizations, and other groups battling to preserve their jobs and communities respond to such characterizations of their position? First, they should meet the challenge head-on, by acknowledging that their position is indeed "conservative," if that term is understood to mean that they favor conserving existing communities in both the physical and social sense. They must emphasize the waste of resources inherent in our society's commitment to unrestrained mobility of capital. They must make clear that in opposing disinvestment in their communities, they are in effect supporting the conservation of resources.

Yet opposition to disinvestment in Frostbelt communities is not an adequate platform with which to attract broad support. Except for those whose jobs are actually threatened or whose neighborhoods are actually redlined by banks and insurance companies, it is difficult to excite public interest in an abstract concept like disinvestment. Thus opponents of disinvestment must find ways to dramatize the issues in immediate human terms, as the organizers of this conference tried to do with their dramatic readings. In addition they must put forward some positive alternatives to existing economic institutions, in the form of employee stock ownership plans, cooperatives, and community development corporations.

While none of these alternatives by itself represents a solution to the problem of capital flight, taken together they comprise an imaginative political agenda with a potentially broad appeal.

Notes

1. Fully 38 percent of the 100 attendees who responded to a questionnaire said they got their information on plant closings through personal contact with friends and family members who have been affected by plant closings.

2. E. E. Schattschneider, *The Semi-Sovereign People* (New York: Holt, Rinehart, & Winston, 1960), p. 71.

3. Peter Bachrach and Morton Baratz, *Power and Poverty: Theory and Practice* (New York: Oxford University Press, 1970), p. 18.

4. Joint Economic Committee of Congress, Subcommittee on Fiscal and Intergtovernmental Policy, "Central City Businesses—Plans and Problems," 95th Congress, 2nd session (Washington, D.C.: U.S. Government Printing Office, Jan. 1979), p. 23.

5. Brecker, "Business Loves the Sunbelt," *Fortune*, 1977, pp. 134–36.

6. John Mollenkopf, "Paths Toward the Post Industrial Service City: The Northeast and the Southwest," in R. Burchell and D. Listokin, eds., *Cities Under Stress* (New Brunswick, N.J.: Rutgers University Center for Urban Policy Research, 1981), p. 105.

7. Barry Bluestone and Bennett Harrison, *Capital and Communities: The Causes and Consequences of Private Disinvestment* (Washington, D.C.: The Progressive Alliance, 1980), p. 42.

8. See Arthur Hochner's work reported in Chapter 8 of this volume.

9. For a detailed discussion of the profitability issue, see Bluestone and Harrison, *Capital and Communities*, pp. 199–211.

10. Joel Havemann, Neal Pierce, and Rochelle Stanfield, "Federal Spending: The North's Loss Is the Sunbelt's Gain," *National*

Journal, June 26, 1976, pp. 878–91. See also a follow-up article by Joel Havemann and Rochelle Stanfield, "A Year Later, the Frostbelt Strikes Back," *National Journal*, July 2, 1977, pp. 1028–37.

11. For an extended argument to this effect, see Richard Morris, *Bum Rap on America's Cities: The Real Causes of Urban Decay* (Englewood Cliffs, N.J.: Prentice-Hall, 1980).

12. Mary Fitzpatrick and Peter Tropper, *Tax Cuts for Business: Will They Help Distressed Areas?* (Washington, D.C.: Northeast-Midwest Institute, Sept. 1980), p. 3.

13. Ibid., p. 41.

14. Ibid., p. 24.

15. Roger Vaughan, "Federal Policy and State and Local Fiscal Conditions," in Norman Glickman, ed., *The Urban Impacts of Federal Policies* (Baltimore: Johns Hopkins Press, 1980), pp. 467–93; David Mulkey and B. L. Dillman, "Location and the Effects of State and Local Development Subsidies," *Growth and Change* 7 (1976): 37–43; U.S. Department of Commerce, Economic Development Administration, "Local Business and Employment Retention Strategies" (Washington, D.C.: U.S. Government Printing Office, Sept. 1980).

16. Edward Schwartz, "The Private Sector Solution" (Philadelphia: Institute for the Study of Civic Values, Dec. 1980). Schwartz's finding for Philadelphia is supported by other research on state-level tax abatements. See, for example, Bennett Harrison and Sandra Kanter, "The Great State Robbery," *Working Papers* (Spring 1976), and Bennett Harrison and Sandra Kanter, "The Political Economy of States Job Creation Business Incentives," *Journal of the American Institute of Planners* (Oct. 1978).

17. Neal Pierce, "Should the Cities Knuckle Under to Industry?" *Philadelphia Inquirer*, Nov. 10, 1980, p. 11A.

18. *Report of the President's Commission for a National Agenda for the 80s* (New York: New American Library, 1981).

19. Harvey Molotch, "The City as a Growth Machine: Toward a Political Economy of Place," *American Journal of Sociology* 82, no. 2 (Sept. 1976): 321.

20. Gregory Squires, "Runaway Factories Are Also a Civil Rights Issue," *In These Times*, May 14–20, 1980, pp. 10–11.

21. See Helena Lopata, "The Chicago Woman: A Study of Patterns of Mobility and Transportation," *Signs: Journal of Women in Culture and Society* 5, no. 3 (Spring 1980): 161–69.

22. Douglas Campbell, "Workers Accept Offer from Container Corp.," *Philadelphia Inquirer*, March 3, 1981, p. 1.

23. Cathy Trost, "Firm Ordered to Pay $360,000 for Leaving City," *Detroit Free Press*, Jan. 10, 1981, p. 3A.

24. Jeremy Rifkin and Randy Barber, *The North Shall Rise Again* (Boston: Beacon Press, 1978). See also Carol O'Cleireacain, "Toward Democratic Control of Capital Formation in the U.S.: The Role for Pension Funds," in Nancy Lieber, ed., *Eurosocialism and America: Political Economy for the 1980's* (Philadelphia: Temple University Press, 1982).

25. The Wharton Applied Research Center, *Factors Influencing the Economic Development of Pennsylvania* (Philadelphia: University of Pennsylvania, Sept. 1979).

Chapter 2

Deindustrialization and the Abandonment of Community

Barry Bluestone

One does not have to search very hard these days for statistics that graphically illustrate the depth of the current economic crisis in America. The persistence of double-digit inflation despite double-digit interest rates, the fact that nearly 8.5 million workers, including 17 percent of all non-whites in the labor force, are now unemployed, the existence of a bankruptcy rate among small business that is the highest since 1962, and an annual federal deficit that could easily exceed $145 billion by 1984 are certainly sufficient indicators of how bad things have become. But there is one startling set of statistics that seems to sum up, and at the same time eclipse, all of these. In the single decade ending in 1980, the city of Detroit saw one-fifth (20.5 percent) of its population abandon the community.[1] Over 300,000 fewer people were living in the Motor City at the time of the last census as

Many of the ideas in this paper were developed jointly with my colleague Bennett Harrison in the course of researching our forthcoming book, *Capital vs. Community: The Deindustrializing of America* (New York: Basic Books, 1982 forthcoming).

compared with the one in 1970. To put the issue in perspective, one need only note that since 1950 nearly as many people *left* Detroit (650,000) as live in the nation's 13th-largest city, San Francisco.

What makes this particular statistic so disturbing is that Detroit is not alone. Its collapse is more severe than that of most other communities because of its dependence on the depressed auto industry, but other towns and cities throughout the northeastern and midwestern states and even in parts of the Sunbelt are suffering the same fate. For example, St. Louis, dependent on the auto and aircraft industries, lost more than 27 percent of its residents during the last decade alone. Buffalo, its economy based on auto and steel, lost 23 percent; Cleveland, 24 percent; Atlanta, 15 percent; and Birmingham, Alabama nearly 6 percent. People abandon a community for many reasons, but one of the most important is simply the lack of jobs.

Sudden and permanent job loss is a well-known story in places like Youngstown, Ohio, until 1977 one of the nation's preeminent steel towns. Today the steel industry has all but disappeared from the city. First Youngstown Sheet and Tube and then U.S. Steel abandoned the community in rapid succession. The same kind of corporate flight is occurring in hundreds, if not thousands, of other communities that many of us have never heard of—places like Ontario, California, where General Electric is about to close the nation's oldest steam iron plant; Anaconda, Montana, where ARCO bought up and then closed down the town's huge copper smelter that had been in operation for seventy-five years; and Sheffield, Alabama, where Ford has just ordered the closing of a major foundry. Each of these communities is undergoing what can best be described as "deindustrialization"—a deep and often persistent disinvestment in the community's capital base. The very foundation for economic activity in these communities—the financial capital and the plant and equipment needed for production—is vanishing. In some cases the drain of capital is the result of truly unforeseen circumstances stemming from the day-to-day anarchy of economic competition. In most cases, however, deindustrialization is the result of deliberate and systematic planning by those who are entrusted with making "prudent" investment decisions for private capital.

In the wake of disinvestment, people are being forced in ever larger numbers to abandon their communities, seeking not so much greener pastures elsewhere, as ones that are not as economically parched as those they are forced to leave. In the course of this process, they are forfeiting something quite precious—their sense of security and their desire for community. What is happening to America—and why?

The Extent of Capital Mobility in the U.S.

Despite daily newspaper reports on plant closings and other forms of capital disinvestment, we have very little information about the total number of closings or the number of jobs affected by them. The government keeps excellent statistics about our personal lives, including detailed counts on everything from our annual incomes to the number of bathrooms and air conditioners in our homes. Yet in its zealous pursuit of knowledge about personal hygiene and our summer comfort, it collects virtually no statistics on the number of new plants opened each year in our communities or the number that close down. We know practically nothing from government data about General Motor's movement of capital from one place to another, or even the possibility that it is entirely abandoning the United States.

To fill this void, we have been forced to rely on data collected by the Dun & Bradstreet Corporation. Dun & Bradstreet is in the business of rating the performance of individual companies so that private investors and financial institutions can make more intelligent decisions about their own investments. D & B keeps track of firms which in 1976 were responsible for employing about half of all non-government workers in the nation. The corporation counts new establishment births, establishment deaths, and the interregional migration of individual company units. The data cover not only manufacturing, but the extraction industries, construction, retail trade, and services.[2] Using these data, we have been able to estimate the extent to which new jobs were created by the start-up of new establishments and the extent to which jobs were destroyed by disinvestment in existing ones.

Table 2-1 presents our findings on employment creation and destruction in the private sector from the end of 1969 through the end of 1976, the period for which the D & B data were available. As the table indicates, the sheer amount of employment turnover reflecting capital investment, disinvestment, and mobility is extraordinary. In the country as a whole, between 1969 and 1976, private investment in new business plants (including stores, shops, warehouses, and offices, etc.) created about 35 million jobs, an increase of 50 percent over the 1969 base. This amounts to an average of 5 million jobs created each year as the direct result of plant openings. Presumably this is all to the good.

The bad news is that, by 1976, shutdowns of existing facilities had wiped out 44 percent of the jobs that had existed in the country back in 1969. A total of *31 million* jobs were eliminated by plant closings over the seven year period. This

Table 2-1. Jobs Created and Destroyed by Openings, Closings, Relocations, Expansions, and Contractions of Private Business Establishments in the United States, by Region, 1969–76

Region	*In Thousands of Jobs*								
		Employment Change, 1969–76							*D&B Sample as a Proportion of U.S. Dept. of Labor Estimate of Total Private Employment**
		Jobs Created		*Jobs Destroyed*		*Net Job Change*		*Ratio of Jobs Destroyed by Closings to Jobs Created by Openings*	
	No of jobs in 1969	*By openings and inmigrations**	*Expansions†*	*By closures and outmigrations‡*	*Contractions†*	*No.§*	*Pct. ‖*		
U.S. as a whole	70,324.3	34,798.0	26,074.8	30,820.5	18,457.3	11,595.1	2.2	0.89	.48
Frostbelt	38,758.1	15,288.7	12,518.2	15,237.7	9,577.1	2,992.1	1.1	1.00	.51
Northeast	18,669.8	6,480.6	5,641.1	7,719.1	4,701.6	−299.1	−0.2	1.18	.52
New England	4,542.0	1,585.4	1,439.5	1,820.8	1,210.3	−6.2	−0.0	1.15	.56
Mid-Atlantic	14,127.8	4,895.2	4,201.6	5,898.3	3,491.3	−292.8	−0.2	1.19	.51
Midwest	20,088.3	8,808.1	6,877.2	7,518.6	4,875.5	3,291.2	2.3	0.85	.50
East North Central	14,756.1	6,282.1	4,711.1	5,340.2	3,514.4	2,138.6	2.1	0.85	.52
West North Central	5,332.2	2,526.0	2,166.1	2,178.4	1,361.1	1,152.5	3.0	0.86	.45
Sunbelt	31,566.3	19,509.3	13,556.6	15,582.7	8,880.1	8,603.0	3.6	0.80	.45
South	19,867.8	12,227.0	8,376.1	9,701.4	5,415.7	5,486.0	3.6	0.79	.47
South Atlantic	10,180.8	6,199.9	4,073.9	5,079.0	2,889.9	2,305.0	3.0	0.81	.45
East South Central	3,775.8	2,043.9	1,460.2	1,649.5	849.3	1,005.3	3.6	0.81	.52

West South Central	5,911.1	3,983.2	2,842.0	2,972.9	1,676.5	2,175.7	4.7	0.75	.46
West	11,698.5	7,282.3	5,180.5	5,881.3	3,464.4	3,117.0	3.5	0.81	.42
Mountain	2,571.5	1,826.4	1,393.0	1,446.0	716.5	1,057.1	5.1	0.80	.39
Pacific	9,126.9	5,455.9	3,787.4	4,435.4	2,747.9	2,059.9	3.1	0.82	.43

*We aggregate openings (or start-ups, or what Birch calls "births") and "inmigrations" (plants which are new to the area but which are known to have previously existed somewhere else) into a single category. See Birch, app. A., n.(b) for an explanation.

†These columns refer to employment change in establishments that neither relocate nor shut down during the period of analysis.

‡We aggregate "closures" (or shutdowns, or what Birch calls "deaths") and "outmigrations" (plants which previously operated in the area, closed there, and then reopened elsewhere) into a single category. See Birch, app. A., n. (b) for an explanation.

§Because employment change associated with recorded relocations (inmigrations and outmigrations) is so small—between 0.2 percent and 2.0 percent of net employment change over any particular period of time in any state—there is some, but probably little double-counting in these regional and national totals.

‖ This is *not* measured as (openings plus inmigrations plus expansions minus closings minus outmigrations minus contractions) divided by 1969 employment. The reason is that the D & B data, as reformatted by Birch, are organized into three panels: 1969–72, 1972–74, and 1974–76. The sample size increases with each panel. Therefore, we compute a weighted average of the net percentage changes in each panel, using the respective base year employment levels (1969, 1972, 1974) as weights. See Birch, app. A., n. (e) for details.

SOURCE: David L. Birch, "The Job Generation Process," M.I.T. Program on Neighborhood and Regional Change, 1979, app. A. The regional data shown here are sums of the figures for each state; the latter are presented in the Appendix Table at the end of this chapter. The basic unit of observation is the "establishment" (plant, store, or shop). These estimates are based on a non-probability sample of Dun & Bradstreet reports from the Dun's Identifiers File. We inflated the D & B counts of openings, closings, etc. by the ratio of U.S. Dept. of Labor estimates of state by state employment in 1969 to the D & B counts for 1969. Sample size varied from a low of .30 in Alaska to a high of .61 in Rhode Island. (The District of Columbia sample is really an outlier; because of the predominance of government as an employer in D.C., Dun & Bradstreet makes little effort to follow activity there.) See Birch, app. A, n. (a) for details on our estimation procedure.

amounts to about 4.4 million jobs destroyed each year, on average. Knowing what was happening in the Northeast—the final gasps of the mill-based industries in New England and the fiscal crises in cities like New York—one is not surprised to see that the number of jobs destroyed in this area of the country exceeded the number created. One might be surprised by the apparent better employment record in the Midwest, but this is explained by the fact that the major auto, steel, and tire plant closings in Michigan and Ohio did not take place until *after* 1976.

The real surprise is in the South. We know that the overall pace of economic growth has been greater in the sixteen states making up this region than anywhere else. But even in spite of the region's legendary "good business climate," between 1969 and 1976 industry siphoned out of the region enough capital to destroy almost 10 million southern jobs as a direct result of shutdowns, with another 5.4 million lost through cutbacks in existing operations. Some of these "booming" Sunbelt states actually lost more jobs through plant closings than they gained through new openings. Delaware and North Carolina were two states that suffered like those in the Northeast.

Of course, behind these employment statistics lies the real story—disinvestment through plant closings and runaway shops. David Birch of M.I.T., the first to carefully mine the D & B data, has calculated the amount of actual plant closing activity during the 1969 to 1976 period.[3] His efforts uncover a rate of establishment closings much higher than previously believed. As expected, he finds very high turnover among small, mostly "mom and pop" establishments, especially in retail trade and services. Overall, almost three-fifths of all establishments with twenty or fewer employees that were operating in 1969 in a given state were out of business (or had moved to another state) by 1976. What was surprising, however, was the rate at which even large manufacturing establishments ceased operating. Nationwide, *30 percent* of all manufacturing plants with one hundred or more employees closed down during that seven year period. Again, the greatest surprise was in the South, which had the highest rate of plant closings, 34 percent, compared with 31 percent in the Northeast, 25 percent in the North Central states, and 30 percent in the West.

As we shall suggest later in this paper, very few of these plant closings were due to bankruptcy. The more common variety of plant closing occurs when a multi-plant enterprise shuts down operations in one of its establishments in order to make resources available for investment in another region (including abroad) or in some other industrial sector of the economy. From the firms's perspective, this may the most profitable thing to do. But from the worker's and the general community's point of view, this strategy is often disastrous.

The Personal and Social Costs Left Behind

With so many plant closings, there is bound to be widespread unemployment. Yet the social costs of deindustrialization go well beyond joblessness *per se*. Detroit, first struck by the dispersal of auto production from the central city shortly after World War II, and then clobbered by auto imports, high interest rates, and national recession, has an unemployment rate that now affects nearly a quarter of the city's labor force. Moreover, in many cases unemployment does not disappear very readily. For example, when *Fortune* magazine went into Youngstown, Ohio two years after the Lykes corporation shut down its acquisition, Youngstown Sheet and Tube, its reporter found that 15 percent of the 4,100 directly affected workers were still jobless, 35 percent had been forced into premature retirement, and 20–40 percent had taken jobs that involved substantial wage cuts.[4]

Indeed, the earnings losses from shutdowns are more often than not substantial and long-lasting. Studies by Louis Jacobson, a researcher at a division of the Center for Naval Analysis, indicate that in the first two years after being displaced from the auto industry, workers on average lose 43.4 percent of their expected earnings.[5] Even after *six years*, workers are earning nearly 16 percent less than if they had been able to retain their auto industry jobs. In other industries similar long-term earnings losses are found: 18.1 percent in meatpacking, 14.8 percent in aerospace, and 12.6 percent in steel.

Along with the loss of their jobs, workers often find themselves without their health insurance and portions of their pensions. In a significant number of cases, banks foreclose on their mortgages. It is no wonder, then, that persistent high unemployment, particularly in communities suffering from complete deindustrialization, is associated with a range of serious social "pathologies." M. Harvey Brenner, the leading expert in this field, calculates that every one percentage point increase in the national unemployment rate leads to 37,000 total additional deaths, 920 suicides, 650 homicides, 500 deaths due to cirrhosis of the liver, 4,000 state mental hospital admissions, and 3,300 additional state prison admissions.[6] Other studies corroborate the toll that deindustrialization takes on its victims.[7]

The social damage goes well beyond individual families. Entire communities are disrupted. Tax revenues decline at the same time that the need for public services rise. Once again, Youngstown provides an excellent example. In the wake of Youngstown Sheet and Tube closing, Policy Management Associates was employed to make a forecast of the

Barbara Baker: Campaign for Human Development, U.S. Catholic Conference—used with permission

Children of Eaton workers supported by members of the Amalgamated Clothing and Textile Workers Union, May 18, 1981.

probable impact on tax revenues in Campbell township where the plant was located. They concluded that, in the first thirty-nine months following the shutdown, the communities around Youngstown would lose up to $8 million in taxes; the county government would lose another $1 million; the state $8 million; and the federal treasury as much as $15 million; for a total tax loss that could be as high as $32 million.[8] As it turned out the losses to the community's coffers were so great that the townspeople were forced to increase property taxes by more than $11 million in one year, from $39.8 to $51 million. Even then, the Campbell schools had to obtain a $750,000 loan from the state to keep classrooms open. The school district became the first in Ohio to apply for a second emergency loan from the state to avoid total bankruptcy. Unable to fund its schools, the area has also had to cut back spending on police and fire protection and the remaining public services that residents had come to depend upon. Private disinvestment thus sets in motion a dynamic that ultimately leads to the shutdown of the entire community. Unable to cope, people finally abandon ship.

A Nation of Migrants

For those who can no longer make a living in the deindustrialized regions or sectors of the economy, the standard prescription has always been: "move." And this is precisely what has happened. America has become a nation of industrial refugees. Between 1970 and 1979 nearly 7 million people moved to the South from other regions of the country. Another 4.7 million moved to the West.[9] This stream of migrants is so vast that if they had all come from the six New England states, this entire census region would have been left without a single resident. Of course, there was outmigration from the Sunbelt and return migration to areas like New England, in part because of the massive amount of disinvestment going on in some areas of the South. Still, the South and West combined had a *net* gain from population migration of nearly 3.7 million residents during the 1970s.

What is more, the pace of migration during the second half of the decade actually exceeded that of the first, as shown in Table 2-2.

Table 2-2. Regional Migration Patterns, 1970–74 and 1975–79

	Numbers in Thousands			
	North-east	*North Central*	*South*	*West*
1970–74				
Inmigration	1,035	1,800	3,377	2,141
Outmigration	1,993	2,512	2,312	1,536
Net migration	−958	−712	+1,065	+605
1975–79				
Inmigration	1,035	1,830	3,585	2,552
Outmigration	2,138	2,737	2,513	1,615
Net migration	−1,103	−907	+1,072	+937

SOURCE: U.S. Department of Commerce, Bureau of the Census, *Current Population Reports*, "Geographical Mobility: March 1975 to March 1979," Series P-20, No. 353, Table A, p. 1.

This was particularly true for the north central states which witnessed a wave of auto, steel, and tire plant closings that began in 1975 and accelerated after 1978. Inmigration to this region was of equal magnitude in both halves of the decade, while outmigration increased by nearly 9 percent. (For the record, such evidence makes it likely that the number of jobs destroyed by plant closings in the period since 1976 actually exceeds the number lost during the earlier period covered in the D & B sample.) Nationwide, so much migration took place that nearly one in twelve Americans sixteen years of age and older was living in a different state in 1979 from that in which he or she lived just five years earlier.

Not all of this migration reflects workers being "pushed out" of communities that no longer can offer employment opportunity. People move out of boomtowns like Houston every day because of the "pull" of even better opportunities elsewhere. Still, survey evidence indicates that migration is an "inferior" good. Unless there are compelling economic reasons to abandon one's community, people generally prefer to remain where they are. The direct costs of moving, the psychic costs of separating from one's family and friends, and the costs of disposing of assets (e.g., selling a home with a 5 percent mortgage to buy one carrying 18 percent!) ensure that the majority of migrants leave under the duress of unemployment or underemployment rather than as a consequence of attractive economic opportunities elsewhere.[10]

Understanding the Deindustrialization Phenomenon

In the first half of this chapter, we presented employment data that revealed an enormous amount of capital disinvestment in the United States over the past decade. Tens of thousands of establishment closings, involving literally tens of millions of workers, have left communities reeling. Some like Detroit may never recover. In others, we can expect the standard of living to decline, if not in absolute terms, then relative to other communities.

How are we to understand this phenomenon? Economists have attempted to explain capital disinvestment behavior, and more specifically capital mobility, with a battery of hypotheses ranging from standard location theory (it's all a matter of relative costs) to supply-side Reagonomics (it's all a matter of taxes and government red tape!). None of these explanations seem very satisfactory for they either become tautological, as in the relative cost case, or they place excessive blame on rather inconsequential factors as in the supply-side arguments. What is required is an historically valid analysis that accounts for the strategies adopted by corporate managers in making decisions over capital investment. Here we can only sketch such an analysis, leaving to future work the task of filling in the details.

Our analysis suggests that corporate investment strategy since the early 1970's can best be understood as a response to three conditions in the U.S. economy: first, an extraordinary increase in the degree of international competition that seriously undermined profits and the investment process; second, the post-war series of labor victories that successfully constrained the flexibility of management to respond to economic crisis; and third, a transportation and communications revolution that provided a "permissive technological environment" for rapid and extensive capital dispersal. To begin let us consider the impact of labor's victories since the Great Depression.

From the middle of the 1930s to the 1970s, organized labor in the U.S.—as well as in other industrialized nations—won major concessions on a broad set of issues that ultimately limited capital's flexibility in its use of labor. A small indication of this loss in flexibility is found simply in the sheer size of contract documents negotiated between unions and management. The initial agreement between the United Automobile Workers (UAW) and General Motors Corporation covered altogether one and a half pages when it was signed in 1937.

The contract contained just one provision, the recognition of the UAW as the sole bargaining agent for GM's hourly-paid workers. By 1979, the UAW-GM contract, with its extensive array of provisions covering each production unit, contained thousands of pages. In exacting detail the contract specifies hundreds of items from wage scales and a cornucopia of fringe benefits to limits on subcontracting, the pacing of each machine and assembly line, and even establishes some rules governing the introduction of new technology. Each of these rules and regulations was put in place by labor for the explicit purpose of increasing worker job security and limiting the discretion of management. With the important—indeed absolutely critical—exception of limiting the right of corporate management to reduce the aggregate size of its labor force, these incursions of organized labor were highly successful. On a whole range of matters covering production, labor could bind the hands of management when the latter attempted changes that threatened job security. Jurisdictional lines between trades were enforced. Seniority provisions prevented management from choosing between workers for promotion or layoff. Health and safety rules forced management to design production processes that were safe, often regardless of cost. And even the speed of the assembly line became a matter for negotiation. As long as management was forced to deal with labor where workers were well organized, it was constrained to operate within the set of rules that unions had long struggled to secure.

Moreover, using the power of the state, labor won important concessions from industry through the regulatory process. Minimum wages, fair labor standards, occupational health and safety provisions, equal employment opportunity, extended unemployment benefits, and improvements in worker's compensation comprise only a partial list of the gains made by labor during this period. Taken together, these victories limited management's ability to extract the last ounce of productivity from labor and thus boost profits.

During the heyday of monopoly capital, roughly 1941–1971 (ending symbolically with America being forced to abandon its fixed exchange rate), industrialists were able to reap healthy profits while affording these concessions to organized labor. But with the rebuilding of the European and Japanese economies, global competition forced U.S. corporate management to find fresh ways to circumvent union rules and hold the line on wages in an attempt to reverse declines in profit rates in key sectors of the economy. In one major industry after another—steel, automobile, rubber—labor-management negotiations have taken a dramatic turn: the corporations now are asking for concessions from labor, not the other way around. Ford, for example, recently demanded a 50 percent wage cut as a requirement to keep its Sheffield, Alabama foundry open. (The union refused.) These demands go beyond wage rollbacks; they involve concessions on the introduction of new technology, work pacing, and regulations guarding worker health and safety.

Historically, employers have continually changed their labor market strategies to meet the challenge to profitability waged by workers. The movement from entrepreneurial to technical to bureaucratic control documented by Edwards in *Contested Terrain*, and attempts to control workers through the use of welfare capitalism, Taylorism, Fordism, and even Quality Circles, reflect management's best efforts to extract greater productivity from labor.[11] The problem is that the higher levels of control entail a loss of flexibility in dealing with labor. Bureaucratic control, particularly when it operates within the context of a union shop, transforms the employer's wage bill from a variable to a (quasi-) fixed cost. This rigidity, not only in the wage bill narrowly defined, but more importantly in the control aspects of the labor process, has posed a serious threat to capital struggling to maintain its accustomed levels of profit in the global marketplace.

The obvious response of corporate management, and indeed the goal now being pursued, is to make labor once more a variable cost component in production. To accomplish this goal requires that management disarm organized labor of its standard weapons: the grievance process, various job actions, and work stoppages. What makes the disarming process possible is the enormous increase in the ability of corporate managers to physically relocate production; that is, to move the locus of production or disperse it, and to inexpensively coordinate spatially dispersed production from a central headquarters. Put simply, capital mobility (or merely the threat of disinvestment) has become the most powerful mechanism available to employers for extracting concessions from organized workers and thus reinstituting flexibility in the labor process. Furthermore, it has become the primary weapon in the corporate labor relations arsenal for reversing the regulatory gains won by labor from local and federal government.

The capital mobility option has always been available to some extent. For example, back in the 19th century, the opening of the Erie Canal allowed firms to transfer production to communities all through upstate New York. What is different today is the distance and speed over which that transfer can take place. Satellite-linked telex communications and wide-body jet cargo aircraft provide an environment that, for all practical purposes, allows production to become spatially free. This aspect of the "automation revolution" has much more profound implications for capital/labor relations than even the introduction of computer-controlled tools and versatile factory robots. Satellite communications permit central management to oversee worldwide operations at close to the speed of light, while wide-bodied cargo aircraft permit the movement of physical commodities at near the velocity of sound. Hence the permissive technological environment provides speed, and even more importantly, control. Moreover, this technological revolution promises to provide even faster modes of communication and transportation in the near future. Visitors to the Long Beach McDonnell-Douglas plant find to their amazement that this aircraft company has on its drawing boards a jet freighter for the 1990's capable of flying a million pound payload (or 1,400 passengers) across any continent or ocean. (The only reason it has not been built, says the company guide, is that no airport runway in the world is solid enough to prevent the aircraft from plummeting through the concrete upon touchdown!)

The capital mobility option provided by new technology shifts the fulcrum of bargaining power in favor of capital to an unprecedented degree. It gives employers the ability to insist upon smaller wage improvements in future bargaining, and in a growing number of instances, permits them to effectively demand wage rollbacks. Even more important from management's viewpoint, it provides employers the ability to force labor to accept the introduction of labor-saving technology and the deregulation of job rules. In essence, the capital mobility option provides industry with the power to make "take it or leave it" propositions stick.

Widespread disinvestment, according to this analysis, is therefore hardly the result of old-fashioned mysterious "market forces." Instead it is the outcome of an explicit managerial strategy aimed at raising profits by "disciplining" labor. The discipline is simple: play the game according to our rules, or we leave. In the face of a credible threat to relocate, organized labor is indeed beginning to bend. In one industry after another, labor is making concessions on pay and on issues of shop floor control. But in the meantime, the strategy has resulted in the extraordinary number of plant closings that we noted in the first half of this paper.

It is important to recognize that consistent with this corporate strategy, the vast number of plants being closed (other than very small family-owned businesses) are not shut down as the result of bankruptcy. Instead the phenomenon of disinvestment involves multi-unit firms choosing to close down particular plants or divisions in order to use their cash flow (profits plus depreciation reserves) for more profitable investment elsewhere, or in some cases to move the plant or division to what management believes is a lower-cost area. It is not simply a matter of turning a profit, but making enough of a profit. It is for this reason that our analysis of the Dun & Bradstreet data not only uncovered an immense amount of disinvestment, but showed that firms which are organized as conglomerates had the worst record of jobs destroyed to jobs created. For example, in the metalworking machinery industry in New England, family-owned firms closed down 2.6 jobs for every new job they opened between 1969 and 1976. Corporate-owned firms closed down 1.2 jobs for every new job they created. But conglomerates, like United Technologies Corporation, which owns Pratt & Whitney, the manufacturer of jet engines; Sikorsky, the manufacturer of helicopters; Otis, the producer of elevators; and Carrier, the air conditioner manufacturer, closed down nearly five jobs for every new job it generated in the region. In the supermarket industry, the effect of conglomerates is even more pronounced. Family-owned firms closed down 1.7 jobs for every one they opened; corporations closed down almost none at all—the ratio was only 0.4 to 1.0—but conglomerates eliminated 12.5 jobs for every supermarket job they added in the region.

That bankruptcy is a relatively unimportant factor in capital disinvestment is also supported by statistics compiled by *Fortune* magazine concerning the original Fortune "500" corporate list.[12] To celebrate the list's twenty-fifth anniversary, the editors tracked down the current status of the original Fortune 500. Of the corporations listed in 1955, only 262 were still on the list in 1980. On the surface this seems to indicate a tremendous amount of business failure. But probing below the surface, one finds just the opposite. Of the 238 no longer on the list, 185 were missing because they had been acquired by other Fortune 500 companies and therefore were no longer listed separately. Among the remaining 53 firms no longer on the list, only four were no longer in business at all. And of the four, only *one* had gone bankrupt. The others had closed for "non-business" reasons.

The conclusion we come to is that most capital disinvestment in the U.S. is the result of conscious corporate planning. Moving capital rapidly between regions and sectors of the economy has become the chief mechanism used by corporate management to maximize profit. That deindustrialization of certain regions and sectors of the economy results from this activity is simply not part of the management calculus.

What Is to Be Done

We should note for the record that the problem we have explored here is not capital mobility or capital disinvestment *per se*. In any dynamic, growing economy, it is necessary to transfer investment resources between industrial sectors and often between regions. There is simply no good reason to continue to produce buggy whips or hoop skirts merely to provide workers with jobs. The issue is rather one of the proper *velocity* of capital mobility. If it is too slow, economic growth is stymied. But if it is too fast, a massive price is paid in terms of unemployment and community abandonment. The trick is to find the optimal velocity of capital, the rate that maximizes economic growth subject to some limit on allowable social cost. The unrestrained market, operating under the aegis of managerial strategies to maximize profit, tends to produce capital "hypermobility" that ignores social cost altogether. Therefore, if the public interest is to be served, it is necessary to regulate the market in some way.

There are many ways to regulate the speed of capital investment and disinvestment. Some of these involve tax and subsidy policies. Others involve outright threats of government expropriation of capital if management fails to compensate for the social costs it generates when it moves huge amounts of capital. A moderate alternative lies in between these two. This involves so-called plant closing legislation now being introduced in some twenty states across the country.

Briefly, most plant closing bills include three key provisions: first, advance notification of a plant shutdown or major permanent reduction in employment; second, severance pay for the workers adversely affected by capital's decision to relocate investment; and third, community adjustment fees paid by the firm to the community in lieu of local taxes. These fees are used to help maintain crucial public services such as schools and police and fire protection in the immediate period following a major closing.

Such plant closing legislation is not aimed at stopping capital mobility, but at providing the time and resources to workers and the community to adjust to it. The advance notice provision requires that firms give sixty days to two years notification prior to a shutdown so that individual workers can seek other employment. Alternatively, it gives an opportunity for workers or the community to find another buyer for the closed plant or develop a worker-owned enterprise to operate the establishment. Severance pay provides workers with funds to tide them over the period immediately following the shutdown. Adjustment fees can be used by the community either for maintaining public services, or for a "downpayment" on a community buy-out of the closed facility.

Such legislation has worked effectively in several European countries, including Germany, Sweden, and Great Britain. It has been successful in regulating the pace rather than the amount of capital disinvestment so as to cushion the otherwise socially disruptive effects of sudden plant closings. Given the extent of deindustrialization documented in this paper and our desire to maintain our communities rather than abdicate from them, we need to consider such legislation seriously. Plant closing legislation, despite its opposition from much of the business community, should be a central part of the political agenda for the 1980's.

Notes

1. U.S. Department of Commerce, Bureau of the Census, *1980 Census of Population and Housing* (Advance Reports), Series PHC80-P (Spring 1981).

2. For more information on the use (and possible abuse) of the Dun & Bradstreet data, see Barry Bluestone and Bennett Harrison, *Capital and Communities: The Causes and Consequences of Private Disinvestment* (Washington, D.C.: The Progressive Alliance, 1980).

3. See David Birch, "The Job Generation Process" (Cam-

bridge, Mass.: MIT Program on Neighborhood and Regional Change, 1979).

4. Linda Snyder Hayes, "Youngstown Bounces Back," *Fortune* (Dec. 17, 1979). Given the data in the story, it remains a mystery how the magazine decided to use this title for its story.

5. Louis S. Jacobson, "Earnings Losses of Workers Displaced from Manufacturing Industries," in William G. Dewald, ed., *The Impact of International Trade and Investment on Employment* (Washington, D.C.: U.S. Government Printing Office, 1978).

6. M. Harvey Brenner, *Estimating the Social Costs of National Economic Policy: Implications for Mental and Physical Health and Clinical Aggression*, a report prepared for the Joint Economic Committee, U.S. Congress (Washington, D.C.: U.S. Government Printing Office, 1976).

7. See, for example, Stanislaw Kasl and Sidney Cobb, "Blood Pressure Changes in Men Undergoing Job Loss," *Psychometric Medicine* 32 (Jan.–Feb. 1970); Sidney Cobb and Stanislaw Kasl, *Termination: The Consequences of Job Loss* (Public Health Service, Center for Disease Control, National Institute for Occupational Safety and Health, U.S. Department of Health, Education, and Welfare, June 1977); Don Stillman, "The Devastating Impact of Plant Relocations," *Working Papers* (July–Aug. 1978); and Alfred Slote, *Termination: The Closing of Baker Plant* (Indianapolis: Bobbs-Merrill Co., 1969).

8. Cited in David Moberg, "Shuttered Factories—Shattered Communities," *In These Times*, June 27, 1979.

9. U.S. Department of Commerce, Bureau of the Census, "Geographical Mobility: March 1975 to March 1979," *Current Population Reports*, Population Characteristics, Series P-20, No. 353 (Aug. 1980).

10. See Julie DaVanzo, *Why Families Move: A Model of the Geographic Mobility of Married Couples*, R & D Monograph 48 (U.S. Department of Labor, Employment and Training Administration, 1977); J. B. Lansing and E. Mueller, eds., *The Geographic Mobility of Labor* (Survey Research Center, Institute for Social

Research, University of Michigan, 1977); and Michael J. Greenwood, "Research on Internal Migration in the United States: A Survey," *Journal of Economic Literature* 3, no. 2 (June 1975).

11. Richard Edwards, *Contested Terrain: The Transformation of the Workplace in the Twentieth Century* (New York: Basic Books, 1979).

12. Linda Snyder Hayes, "Twenty-Five Years of Change in the Fortune 500," *Fortune*, May 5, 1980. According to the analysis contained here, 262 corporations on the original 1955 Fortune "500" were still on the list in 1980. Of the remaining number, 185 had been absorbed by merger, twenty-nine were now too small to make the list, fourteen were no longer considered "industrials" (i.e. they had been transformed into financial or service institutions), six were now privately held corporations, and four were out of business.

Chapter 3

"Runaway Plants," Capital Mobility, and Black Economic Rights

Gregory D. Squires

> Over the long haul, it is apparent that the laws of supply and demand have exercised a greater influence on the quantitative employment patterns of blacks than have the laws of the land.[1]

Opportunities for black Americans have long been shaped by the overall health of the economy and the uneven impact of economic fluctuations on the U.S. class structure, as well as by the discriminatory mechanisms operative within American society.[2] While there has been much debate over the relative impact of class and race on the position of blacks in the U.S. (a debate which was recently fueled by Wilson's *The Declining Significance of Race*)[3] there has never been any doubt that blacks have been victimized by both class and race discrimination.[4] The phenomenon popularly referred to as the "runaway plant" and the dynamics of capital mobility in general, though not generally viewed as racial issues, have hammered home in recent years the interconnections between economic rights in general and civil rights in particular. The increasing occurrences of shutdowns and plant relocations have caused many workers, researchers, and even some corporate executives to reconsider fundamental beliefs about the nature of American society and the rights of various groups within that society. Progressive public policy proposals, formerly considered as little more than fodder for academic debates, are now put forth as solutions for all-too-real problems by union leaders and elected officials.[5] In turn, perhaps, inadvertently, these developments have exposed the severe limitations of traditional civil rights tools and suggested some potentially powerful strategies for a more effective black liberation struggle.

This report examines patterns of uneven development which have characterized the American economy since World War II and the ways in which black workers and communities have been affected. After a brief review of the direction of economic development, the impact on communities and individuals affected, and the factors which account for the nature of capital mobility in the U.S., the implications for racial minorities are explored. In addition to documenting what have been generally adverse consequences for the nonwhite community, the legality of management decisions to shutdown, relocate, or otherwise allocate private business activity is examined in light of current civil rights laws and affirmative action requirements. Innovative policy recommendations to combat the adverse effects of capital mobility in general and the racially discriminatory implications in particular are explored. Recommendations are offered to further the objectives of economic democracy throughout the American occupational structure and equal opportunity for the minority population.

Flight to the Suburbs, the Sunbelt, and Beyond

For several decades the U.S. has experienced a flow of people, jobs, investment capital and other resources from central cities to their suburban rings, from older industrialized metropolitan areas in the Northeast and Midwest to sunbelt states in the South and Southwest, and from every region of the country to overseas locations. The relocation of a plant from one community to another represents just part of this overall pattern of development, though it is the kind of event which has grabbed most of the headlines and has in fact created severe hardships for many communities and families. Equally important, if not more so, in terms of the overall patterns of uneven regional development, are the rates of expansion and contraction of existing firms and of the birth of new firms.[6] Precise data on the number of plant closings, relocations, expansions, and births are not currently available, though recent studies of New England states indicate progress is being made in overcoming this limitation.[7] Scattered data which are available, in conjunction with census and other demographic information permit at least a general understanding of the directions and consequences of economic development throughout the nation. A few basic statistics illustrate the overall pattern.

Between 1960 and 1977 the nation's central city population increased by just 3.5 percent (from 58 to 60 million) while the suburban population grew by 51 percent (from 55 to 83 million).[8] Changes in the labor force have been even more dramatic. During those years the central city workforce rose 17 percent (from 23 to 27 million) compared to an increase of 95 percent (from 20 to 39 million) in the suburbs. And the suburban jobs are paying better. Between 1960 and 1976 the median family income for central city residents increased by 88 percent (from $7,417 to $13,952) compared to an increase of 109 percent (from $8,351 to $17,440) for suburban families.[9] The Chicago metropolitan area typifies shifting business patterns. Between 1966 and 1976 the city experienced a 16 percent net decrease of manufacturing firms while the suburbs enjoyed a 41 percent growth.[10] And between 1970 and 1978 the city lost 15 percent of its retail establishments while the suburbs gained 33 percent.[11]

During roughly the same period of time (1960 to 1975) states in the Northeast and Midwest had a population gain of just 9 percent (from 97 to 106 million) compared to a 24 percent increase (from 55 to 68 million) in the South.[12] Similarly, the civilian labor force in the northeastern quadrant (northeastern and midwestern states) rose 26 percent (from 38 to 48 million) compared to 50 percent (from 20 to 30 million) in the South.[13] Manufacturing employment declined by 9.9 percent in the New England states and 13.7 percent in the Midwest while it increased by 43.3 percent in the South.[14] And family income increased faster in the South than in the northeast quadrant; by 179 percent (from $4,465 to $12,443) in the South, by 151 percent (from $5,892 to $14,813) in the Midwest, and by 144 percent (from $6,191 to $15,085) in the Northeast.[15] In addition, capital investment for new manufacturing equipment increased by 60 percent in the Northeast, 178 percent in the Midwest, and 349 percent in the South between 1967 and 1978.[16]

Jobs and capital are also moving overseas. According to a State Department study, the U.S. experienced a net loss of 1.06 million jobs because of corporate movement to foreign countries between 1966 and 1973.[17] And between 1950 and 1974 American investments overseas increased from $11.8 billion to $118.6 billion.[18]

One consequence of these developments has been the increasing concentration of wealth in fewer hands. According to one Senate subcommittee report, foreign investment by

American multinationals resulted in a 4 percent drop in labor's share of national income.[19] In light of other evidence which shows that the richest one percent owned almost 26 percent of the nation's wealth[20] and that, according to another congressional study, "for all practical purposes, the 5.2 percent of the adult population who own two-thirds of the value of all privately held corporate stock have a large measure of effective control over virtually all corporate assets,"[21] the Federal Trade Commission concluded that recent economic developments "pose a serious threat to America's democratic and social institutions by creating a degree of centralized private decision-making that is incompatible with a free enterprise system."[22]

Aggregate statistics, however, do not reveal the human tragedy that often accompanies the shutdown of a community's major employer or the gradual reduction of economic activity in a city or region. When a plant closes, or cuts back production, a number of adverse trends are set in motion. First, of course, many workers lose their jobs and their families lose their source of financial support. In addition, suppliers and customers of the plant must cut back or seek new markets. Tax revenues lessen precisely at a time when demands for public assistance and services are on the rise. Property values decline. Donations from private charities are cut. As the economic climate deteriorates, local businesses are hesitant to expand and others will take a second look before moving into the community, thus compounding existing problems.[23] Just the threat of a shutdown can cause unions to compromise necessary wage increases or affirmative action commitments and communities to forego safety and environmental protection measures.[24]

Ultimately, of course, it is individuals who suffer. Divorce, depression, wife and child abuse, alcoholism, and several

Barbara Baker: Campaign for Human Development, U.S. Catholic Conference—used with permission

Al Tilson, chairman of the Philadelphia Unemployment Project.

other mental and physical illnesses have been associated with plant shutdowns.[25] One study found a suicide rate among workers displaced by shutdowns to be thirty times the norm for the population in general.[26]

Perhaps these various costs would be somewhat tolerated if they could be attributed to the "natural" workings of a free market adjusting itself to a new equilibrium in which society in general would benefit from the resulting maximum efficiency in production and distribution of goods and services. Though important, market factors are not the sole considerations involved in decisions to shut down, relocate, expand, or open new facilities. And frequently when cost considerations do enter in, overriding public needs are sacrificed for the profit of private industry.

Market factors such as the need for space, cost of land, proximity to markets and skilled labor, and the various costs involved in doing business are important considerations in location decisions.[27] Among those cost considerations are workmen's compensation, unemployment insurance, welfare, and taxes.[28] But according to one Fantus Corporation Vice-President, "Labor costs are the big thing, far and away. Nine out of ten times you can hang it on labor costs and unionization."[29] The attraction of the South, therefore, is evident since as of 1970 wages in the South were 20 percent below the national average[30] and just 13 percent of southern workers are unionized compared to 25 percent nationally.[31] And in Taiwan unskilled laborers earned $2.70 per day in 1975.[32]

The conglomerate-merger movement of the late 1960's and early 1970's, stimulated by Federal tax policies unrelated to efficiency or social utility resulted in the shutdown of many profitable operations.[33] In several cases, viable subsidiaries of some conglomerates were shut down because their profit margins did not meet arbitrary levels required by the parent corporation, occasionally in instances where the requirement was three or four times greater than what the subsidiaries' competition was earning.[34]

In other cases, simple bad management accompanied by unsound and occasionally illegal accounting practices has contributed to shutdowns. The now infamous shutdown of the Campbell Works of Youngstown, Ohio's Sheet and Tube Company by its parent, the Lykes Corporation, is a case in point. The company cited foreign competition, environmental regulation, and government price restraints as the principal causes of the closing. Further research demonstrated that because of the high debt Lykes had incurred to purchase the plant, it was forced to utilize the profits from the Campbell works to pay its creditors and was unable to make the appropriate investments to modernize the steel company to keep it competitive.[35] This shutdown was not the result of natural market forces, unexpected developments in the area of foreign competition or government regulation. A Justice Department study issued before the takeover by Lykes advised against the merger precisely because it was anticipated that the debt assumed by Lykes would make it unable to finance the necessary modernization. The study predicted that the plant would shutdown in ten years. The prediction was off by just one year.[36]

A number of factors unrelated or only tangentially related to market considerations also contribute to the uneven development experienced by the U.S. in recent decades. Fear of crime and a preference to avoid the congestion of many urban areas often contribute to relocations from central city to suburban locations.[37] Lifestyle preferences of corporate executives play a role.[38] Perhaps the most important non-market factor, however, is government.

Through its taxation, expenditure, employment, and law enforcement practices government at all levels has contributed to the shape of economic development in the U.S. For example, northeastern and midwestern states receive an average of eighty-one cents for each tax dollar they send to Washington while southern states receive $1.25.[39] No single step opened the nation's suburbs as quickly as the post World War II highway program.[40] Between 1966 and 1976 Federal civilian employment in central cities dropped by 41,419 while it increased by 26,559 in Standard Metropolitan Statistical Areas (SMSAs) in general. And between 1963 and 1973 Federal employment grew by 9,000 in the Northeast while southern states picked up more than 118,000 new Federal workers.[41] Foreign tax credits and investment tax credits encourage abandonment of existing facilities, thus expediting the decline of older cities in the Northeast and stimulating development in the South and in foreign countries.[42] And "right-to-work" laws enacted by several southern states hamper union organizing thus keeping wages down and contributing to the flight of capital to the Sunbelt.[43]

The importance of state and local tax abatements, loan guarantees, and other governmentally provided assistance has sparked much controversy in recent years. While cities and states throughout the nation utilize such tactics to lure businesses, there is increasing skepticism, even on the part of industry groups, whether or not they affect location decisions and generate new jobs, or simply provide private industry with a subsidy for actions that would have taken place anyway, thus draining the community of scarce revenues and vital services.

The costs of uneven development and economic dislocation are severe. The causes are many and complex. An additional cost associated with these developments which has generally been ignored is the exacerbation of racial inequality. As with dislocation in general, the racial complications do not result simply from the free play of a benevolent market.[44]

Capital Mobility and Equal Opportunity

Several dimensions of the "runaway plant" phenomenon adversely affect minority communities. Population dynamics constitute one such factor. While economic activity has been moving from central cities to the suburbs, and from the Northeast to the Sunbelt, the nation's minority population has not moved with the flow to the same extent as has the majority. Between 1960 and 1975 blacks increased their proportion of central city populations (from 16.4 percent to 22.6 percent) while their representation in the suburbs remained virtually unchanged (4.8 percent and 5.0 percent).[45] And during these years the proportion of blacks in the South declined (from 21 to 19 percent) while it increased in the northeast quadrant (from 7 to 9 percent).[46]

No comparable data are available to assess the racial effect of the flight of capital overseas. However, in light of the fact that American investment abroad has reduced labor's share of national income and the fact that minorities are more dependent on wages and salaries for their income than whites, it is reasonable to conclude that these developments also adversely affect minorities. (Minority-owned businesses control 4.4 percent of U.S. businesses, generate 0.7 percent of U.S. business receipts and employ 0.8 percent of all U.S. workers. Wealth for whites is roughly 4.5 times that of blacks compared to a ratio of one-to-five for income. Whereas 80 percent of minority earnings accrue from wages and salaries, the comparable figure for whites is 75 percent.)[47]

The underrepresentation of black workers at the higher rungs of the American occupational structure (the proportion of whites in professional, technical, managerial, and administrative occupations is twice that of nonwhites[48]) contributes to the discriminatory effects of capital mobility. Management level employees are more likely to be given advance warning of any impending relocation (and therefore more time to plan for it), they are more likely to be offered jobs at the new site and to be offered substantial financial assistance in making the move, or to be provided with job hunting assistance if they chose not to relocate.[49] Minorities also suffer from the fact that those industries sustaining the steepest declines in central cities tend to have relatively high concentrations of minority workers while they are underutilized in industries on the rise,[50] and those which employ relatively large numbers of blue-collar workers (a disproportionate share of whom are minorities[51]) have decentralized at a faster rate than others.[52]

But the discriminatory effects of plant relocations and uneven development are not simply unintended manifestations of discriminatory mechanisms operating in other sectors of society. Private decisions and public policy have frequently been made with the objective, at least in part, of perpetuating racial discrimination. A study of plant relocation by the Equal Employment Opportunity Commission (EEOC) elicited the following statement from one employer who had moved from a central city to a suburban location, "We could not get employees to work on the fringe of the Negro section where we were located."[53] Several others offered similar concerns with inner-city locations, frequently citing the "better class of employees" or the "more stable group of people" in the suburbs. A former manager of a Detroit employer who relocated to a nearby suburb stated that one objective was to "escape the obligation of hiring blacks."[54] And the suburbanization of the nation's population occurred with major support from the Federal Housing Administration which for years stated in its underwriting manual, "it is necessary that properties shall continue to be occupied by the same racial classes."[55]

Some recent research suggests that residential location does not substantially affect black wage levels.[56] But there is abundant survey and case study evidence which indicates that the concentration of minorities in central cities and discrimination in suburban housing markets, coupled with the suburbanization of employment has restricted knowledge of job openings, applications for positions, and hiring within minority communities, and has increased the costs (particularly commuting) of holding a job for minority workers.[57]

The racially discriminatory implications of capital mobility is most explicitly demonstrated in the employment patterns of companies involved in shutdowns, relocations, or expansions. Data for Illinois companies involved in the following types of developments between 1975 and 1978 were obtained from EEO-1 reports for those years:[58]

- shutdown of operations;
- relocation from a central city to that city's suburban ring;
- relocation from a suburban ring to the respective central city;
- relocation from Illinois to another state;
- relocation from another state to Illinois;
- Illinois-based companies which opened up new branches in Illinois and those which branched out in other states; and
- companies based in other states which opened up new branches in Illinois.

Certain limitations of the data must be kept in mind. First, not every employer is required to file EEO-1 reports, and some who are required by law to report fail to do so. For these years approximately half the private sector workforce nationally was covered by reports which were filed.[59] Secondly, of course, this analysis is limited to basically one state and to a three-year time frame, though the phenomenon under investigation is nationwide (if not worldwide) and has been occurring for decades. Third, the birth of new firms is not included because of cost considerations. Fourth, though a shutdown, relocation, or expansion may occur in a given year, such events generally represent the culmination of developments which have been occurring over time. For example, a corporation may gradually invest its profits from one location in another plant located elsewhere so that by the time the former operation is closed few people may be affected or surprised.[60] This ongoing shift of resources is not examined from this type of analysis, although discrete results are.

And finally, a snapshot may underestimate the racial effects of these developments because those effects result from events before and after the actual move. There is evidence, for example, that a disproportionate number of minorities will leave a company prior to the relocation in anticipation of the difficulties it will create. Similarly, a relatively higher proportion of minorities quit shortly after the move when they find the hardships greater than anticipated.[61] Minorities who remain with the company tend to encounter higher commuter costs which, because of housing discrimination in suburban communities, are not simply voluntary expenses.[62] And among workers who lose their jobs, minorities encounter greater difficulties in securing new positions, as indicated by the fact that the average duration of unemployment for whites during the fourth quarter of 1979 was just five weeks compared to 7.1 weeks for nonwhites.[63]

Economic dislocation has affected all groups in Illinois, but minority groups have been particularly hard hit. For example, the 2,380 firms in this sample which shut down between 1975 and 1978 formerly employed 365,032 workers of all races, but 20.0 percent were minorities compared to a statewide minority workforce of just 14.1 percent.[64] Among those firms which relocated from an Illinois central city to the suburban ring, total employment declined from 18,653 to 16,335 with black employment dropping from 23.7 percent to 20.5 percent of the total. In other words, black employment declined by 24.3 percent compared to just 9.8 percent for white workers. Hispanics, interestingly, increased their share from 3.9 to 4.7 percent with the move to the suburbs. Relocation from suburbs to centers of cities resulted in an increase in minority employment from 13.2 percent to 22.7 percent, but less than 800 jobs were involved, hardly compensating for the exodus from cities.

In firms relocating from Illinois to the South, minority employment dropped from 23.3 percent to 21.1 percent, as it did in those firms going from the South to Illinois, from 29.9 to 29.3 percent. Just under 2,000 jobs were involved in the move south, compared to just over 800 in the relocation to Illinois. However, for all firms relocating between Illinois and each region of the country, accounting for approximately 9,000 jobs each way, minority employment increased from 16.2 percent to 17.7 percent for those leaving the state, and from 10.8 to 18.3 for those entering. Relocation, whether within a metropolitan area or from one region to another, is disruptive for a company's employees even if no jobs are lost and each worker stays with the employer. Among each group of firms in this sample, minorities, prior to relocation, were employed in greater proportions than they were represented in the respective regional labor markets, except for those firms which left the Northeast. Minorities experienced a disproportionate share of the disruption created by corporate relocation.

Expansion, or the creation of new branches, increased job opportunities for minorities both absolutely and relatively to the majority population. In each region, new branch offices employed minorities in higher proportions than they are represented in the respective regional labor markets. In Illinois, however, the 182,049 jobs created by such expansions did not offset the 365,032 jobs lost through shutdowns, and minority employment in those new offices was of lower (17.7) percent than in the firms shut down (20.0 percent).

The discriminatory effects of shutdowns, relocations, and capital mobility are all quite real. Given their concentration in those communities experiencing economic decline or relatively lower rates of growth, their concentration at the lower rungs of the occupational ladder and within declining industries, recent public and private spending and employment practices favoring suburban and sunbelt locations, and in some cases, the blatantly discriminatory nature of private sector hiring, location decisionmaking, and public policies, it is clear that the adverse impact of these developments on black Americans is hardly fortuitous. But are they legal?

Legal Implications

To date there is no law, regulation, or court decision (though there are cases pending[65]) which conclusively and expressly address the legal implications, from a civil rights perspective, of plant relocation. However, several civil rights attorneys, including officials with various enforcement agencies, have reached some consensus on at least certain circumstances where a plant relocation would constitute a violation of various federal civil rights requirements. Most of these discussions focus on relocations from predominantly minority or racially-mixed center-city locations to predominantly white suburban areas. The same principles can apply, however, when the relocation is from one state or region to another, or when a corporation opens up a new facility.

Arguably, the relocation of an employer's facilities from an urban to a suburban location where minority workers do not reside violates Title VII of the Civil Rights Act of 1964 unless that employer takes steps to assure equal employment opportunity.[66] Section 703(a) of Title VII, as amended by the Equal Employment Act of 1972 states:

> It shall be an unlawful practice for an employer—
>
> (1) to fail or refuse to hire or to discharge any individual, or otherwise to discriminate against any individual with respect to his compensation, terms, conditions, or privileges of employment, because of such individual's race, color, religion, sex, or national origin; or
>
> (2) to limit, segregate, or classify his employees in any way which would deprive or tend to deprive any individual of employment opportunities or otherwise adversely affect his status as an employee, because of such individual's race, color, religion, sex, or national origin.[67]

A memorandum prepared by the EEOC Office of General Counsel concluded that the transfer of an employer's facilities constitutes a *prima facie* violation of Title VII if:

> (1) the community from which an employer moves has a higher percentage of minority workers than the community to which he moves, or (2) the transfer affects the employment situation of the employer's minority workers more adversely than it affects his remaining workers, and (3) the employer fails to take measures to correct such disparate effect.[68]

Such a transfer adversely affects minority employment in two respects. First, it diminishes the number of minority job candidates in the pool of labor from which the employer would be expected to recruit in future hiring. Second, the transfer would adversely affect the employment opportunities of the firm's current minority workforce because they would face greater difficulties in relocating or commuting to the new location.

A corporate relocation from a central urban to suburban location, or from one region of the country to another, which has a disparate impact on minority employment, therefore, creates a *prima facie* violation of Title VII regardless of the employer's intentions regarding equal employment opportunity. As the Supreme Court ruled in the *Griggs*[69] decision:

> Under the (Civil Rights) Act, practices, procedures, . . . neutral on their face, and even neutral in terms of intent, cannot be maintained if they operate to freeze the status quo of prior discriminatory employment practices. . . . Congress directed the thrust of the Act to the consequences of employment practices, not simply the motivation.[70]

The burden is on the employer to demonstrate an overriding business necessity to justify a seemingly neutral practice, like a relocation, which has a disparate impact on minorities. "Business necessity" has been interpreted narrowly by the courts. An employer must demonstrate that the practice in question is vital to the operation of the business, that it is necessary for the safe and efficient operation of the business, and that it effects an essential competitive advantage. The practice must effectively carry out the business purpose allegedly served, and there must be no available alternative which would better accomplish the business objective or would accomplish it as well but with a lesser differential impact on minorities.

Even where a corporate relocation which adversely affects minority employment meets these stringent "business necessity" requirements, the move may still violate Title VII if the employer has not taken appropriate steps to minimize that disparate impact. An employer has a duty of fair recruitment which requires special efforts when normal recruiting practices would adversely affect minority employment. Corporate relocations represent a case in point. Noting the complex planning process involved in considering and then preparing for a relocation, Rutgers University Law Professor Alfred W. Blumrosen argues, "an employer of significant stature who utilizes sophisticated planning methods could not successfully contend that he was simply unaware that there were any racial or ethnic implications to his move."[71] Once aware of these implications, Blumrosen states, the employer has a legal obligation to "plan for fair employment," And he concluded, "There is no 'business justification' for the failure of a large employer to consider the probable racial impact of a move to the suburbs."[72]

Blumrosen and others have argued that corporate relocations also frequently violate Executive Order 11246 and other regulations with which Federal contractors are supposed to comply, although no regulations addressing relocation specifically have yet been promulgated.[73] Under Executive Order 11246, federal contractors are required to "take affirmative action to ensure that applicants are employed, and that employees are treated during their employment without regard to their race, color, religion, sex, or national origin."[74] Regulations were subsequently adopted by the Office of Federal Contract Compliance, spelling out the details of the affirmative action requirement. Order No. 4 and Revised Order No. 4 call for federal contractors with fifty or more employees, or contracts worth $50,000 or more, to evaluate minority and female representation in all job categories and develop an affirmative action plan, including numerical goals and timetables, to eliminate any deficiencies which may exist in the utilization of minorities or women. These regulations specify the criteria contractors are to use in evaluating their workforces. These criteria include the size of the minority population and workforce in the surrounding community and relevant labor market, as well as the availability of minorities having the requisite skills in an area in which the employer can reasonably recruit, and the existence of training institutions capable of training people in the requisite skills. The regulations also call for corrective action if the contractor finds, among other factors, "Lack of access to suitable housing inhibits employment of qualified minorities for professional and management positions" or "Lack of suitable transportation (public or private) to the workplace inhibits minority employment."[75]

A corporate relocation may also violate a number of other civil rights requirements, depending on the nature of the personnel policies followed by the employer in the process of closing down the former facility and staffing the new one. For example in the pending case of *Abney* v. *Budd*,[76] plaintiffs allege that the Budd Company intentionally discriminated against minorities by refusing job placement assistance to hourly employees (all of which were black or Mexican) while offering such assistance to salaried employees (most of whom were white), and therefore charged the employer with violating section 1981 of the Civil Rights Act of 1866. Plaintiffs also allege that Budd did not pay men and women equal wages for comparable work and therefore charged the company with violating the Equal Pay Act of 1963.

Plaintiffs in the case of *Bell* v. *Automobile Club of Michigan*[77] allege violations of Title VII and the Civil Rights Act of 1866 when the auto club moved from the city of Detroit to a nearby suburb despite the availability of alternative sites within the city, thus, according to the plaintiffs, invalidating any "business necessity" argument. Among the facts noted by the plaintiffs are the declines in the proportion of black job applicants and black hirings after relocation. For the three years prior to the 1974 move, blacks accounted for 53 percent of all applicants and 39 percent of all hirings compared to 26 percent of applicants and 29 percent of hirees for the three years following the relocation.

Several corporations have taken voluntary action in efforts to mitigate the racially discriminatory impact of relocation. The kinds of efforts include contracting with fair housing centers to provide house-hunting assistance, requiring realtors to sign fair housing pledges before referring employees to them, providing rent subsidies, renting entire apartment complexes and then subletting units to minority employees, and filing fair housing lawsuits on behalf of minority employees. Among the corporations which have taken one or several of these actions are such household names as General Motors, General Electric, Standard Oil, Republic Steel, IBM, Procter and Gamble, Shell Oil, Caterpillar, Illinois Bell, Sears, and Stanford University.[78]

Because voluntary measures are not always sufficient and because of the major expense of litigation, a number of civil rights experts have called for the EEOC and the Office for Federal Contract Compliance Program to develop regulations specifying an employer's obligations to assure equal employment opportunity when relocating or opening up a new facility.[79] The regulations would require employers to maintain detailed records pertaining to location decisions which would be available to appropriate authorities. Prior to any move, an employer would have to file an employee impact statement delineating the effect of the proposed move on minority employment. In those cases where a disparate impact on minorities is anticipated, the regulations would indicate the steps required by the employer to mitigate that impact, including but not limited to: providing transportation services or allowances; providing househunting assistance or directly providing housing; pressing local officials, realtors, and lenders to assure equal opportunity; and considering alternative sites that would be less discriminatory; and reconsidering the move altogether.

These voluntary efforts and proposals for administrative actions have evolved from what can today be considered the traditional affirmative action approach. While such policies may ameliorate some of the adverse racial implications associated with the "runaway plant" phenomenon and the social costs of capital mobility in general, they do not confront directly the underlying causes of economic dislocation and its racial effects. Such affirmative actions must be complemented by more fundamental changes in the way resource allocation decisions are made.

Rights of Capital versus Rights of Minorities

Managerial prerogative and economic efficiency are frequently offered as the key rebuttals to recent efforts to lessen the adverse consequences of capital mobility.[80] Yet as several analysts have noted, the public bears and pays for many social costs which are created but not taken into account by private corporations in their cost-benefit analyses.[81] As a result, these critics contend, it is essential to begin challenging, if not usurping, some of the prerogatives long considered by management to be their exclusive domain. As a study prepared for the FTC recently concluded:

> The private decisions of corporate owners and managers impose costs that affect other businesses, employees, and the community at large. The mental and physical well-being of the community is deleteriously affected; increased stress is placed upon the family; the quality of life in the community can be seriously decreased. Decisions with these wide-ranging effects cannot be viewed solely as private prerogatives; the internal decision-making calculations of the firm do not fully reflect actual costs involved. Public concern and participation is needed to insure that improperly estimated economic gains do not impose major economic costs.[82]

The issue, however, is not simply geographic mobility; from cities to suburbs, from the Northeast to the South, or from the U.S. to foreign countries. The heart of the problem is uneven development itself and the fact that ownership and control of the nation's resources are so heavily concentrated as to constitute what the FTC called a threat to "America's democratic and social institutions." In the words of economists Barry Bluestone and Bennett Harrison:

> Plant closings are real, and the injury to workers and communities when a company closes down is real. But there is much more going on, and far more extensive injury, than 'only' what relates explicitly to relocations. . . . The context within which to understand—and resist—plant closings and relocations is the changing political-economic structure of international capitalism, specifically those changes in law, politics, and technology that have increased the mobility of private capital. Ultimately, unless we are prepared to seize control over these conditions, we will at best only cosmetically treat what is only one part of a much larger problem.[83]

Historically the American response to problems of economic dislocation has been to deal with the effects rather than the causes. Unemployment compensation, job training, economic assistance to depressed areas, and a host of other social services have been designed to aid individuals adversely affected by structural changes of American capitalism.[84] To the extent that efforts are made to shape rather than respond to conditions, they generally take the form of tax subsidies, rebates, and other financial incentives offered in hopes of attracting private capital into a community. Such marginal approaches, again, fail to address the underlying structural determinants of uneven development. They treat the existing institutional context of economic development as a given and attempt to improve the lot of various individuals and groups within that context.[85] This is precisely the major limitation of affirmative action and other traditional civil rights strategies. They are oriented towards redistributing the existing pie within the current structural framework of the economic system. What is called for, however, is the development of new structures in which ownership and control are more widely distributed and greater public control of major economic enterprises to assure that public as well as private needs are met,[86] including assurances of equal opportunity for minorities and women.

Terms like "worker self-management" and "economic democracy" have increasingly become part of the economic development lexicon. The most extensive movement in this direction has been in several western European countries where a combination of government regulations and collective bargaining agreements require in many instances labor representation on corporate boards; pre-notification by management of any proposed relocation or reduction of activity; and negotiations of the terms of a proposed relocation or reduction including provisions of such benefits for workers as rights to jobs and relocation subsidies, retraining, time off to seek new jobs, severance pay, and others.[87] Employee ownership is also a prevalent phenomenon in many European countries, with perhaps the most innovative example being the Mondragon cooperative system in Spain, which consists of sixty-five different firms and over 14,000 members.[88]

Similar initiatives are being launched in the United States. Legislation has been introduced in Congress and in state legislatures of Illinois, Ohio, Massachusetts, New York, New Jersey, Connecticut, Oregon, Pennsylvania, Michigan, and others requiring companies to notify employees several months in advance of any proposed shutdown, reduction, or relocation; provide severance pay to affected employees; and contribute to a community assistance fund to aid affected cities. Many of these proposals also call for financial and other assistance for employee groups to purchase and operate facilities that would otherwise close. Wisconsin has already enacted a pre-notification requirement, Maine requires severance pay to affected workers,[89] and Michigan provides support for employee takeovers.[90] The nomination of UAW President Douglas Fraser to Chrysler's board of directors represents a new direction for U.S. corporations.[91] As Chrysler President Lee Iacocca stated, "If the workers are going to have a voice in their own destiny, they should be represented when these crucial decisions are made."[92] Another new direction in American labor relations is the 1979 United Rubber Workers contract which calls for prenotification of any shutdown, preference in hiring of affected workers at other company locations, pension benefits, and the right to negotiate ways to save the plant or the manner in which it will be closed.[93] Employee ownership, though in widely divergent forms, exists with employees owning a majority interest in at least ninety companies.[94] Over fifty instances of employee ownership resulted from employee purchases of operations scheduled to be closed by former owners.[95] Available research indicates that employee-owned firms are more productive, efficient, and profitable. They pay higher wages and enjoy better labor relations and employee morale than comparable firms in the same industry with conventional ownership structures. These findings are further enhanced the greater the share of equity owned by the employees and the greater their decision-making participation in running the businesses.[96]

Other economic development initiatives which have evolved from a similar philosophical framework include public banks (e.g., Bank of North Dakota, National Consumer Cooperative Bank), public investment corporations (Massachusetts Community Development Finance Corporation, Connecticut Product Development Corporation), and community development corporations (e.g., Harlem Commonwealth Council, Lumni Indian Tribal Enterprise Inc., Roxbury Action Program). Like the initiatives discussed above, they represent economic development efforts which, contrary to conventional approaches, are not designed primarily to lure private capital.

Economic Democracy and Civil Rights

Employee ownership, regulation of plant closings, public investment institutions, and more comprehensive affirmative action requirements all represent efforts to increase the flow of economic resources to people and communities not adequately served by private markets and the welfare state. This seemingly disparate array of approaches to the interrelated problems of uneven development and equal opportunity share three critical underlying tenets which distinguish them from more conventional efforts and, therefore, offer potential for success where the tax abatement game has failed. First, they place primary emphasis on meeting public or social needs directly rather than indirectly as a result of an activity aimed principally at maximizing a financial return of privately invested capital. Second, they encourage broader, more equitable, and more accountable control of economic resources and the concomitant power those resources bring, rather than concentration of resources controlled by increasingly fewer, more distant, and anonymous individuals. And third, they are directed toward achieving equality in fact for racial minorities and women within more democratic social, economic, and political arrangements rather than equal opportunity within a rigid, hierarchical system.

Black Americans are victimized by their location in an economic system in which resources are owned and controlled by a gradually decreasing number of private interests, often at severe public expense, as well as by explicit racially discriminatory mechanisms within that system. Efforts to democratize that structure (i.e., to create greater public control over major investment activities to assure that public needs are adequately served, and to create broader ownership and control within individual enterprises) are civil rights battles. None of this situation, however, minimizes the importance of effective enforcement of current equal opportunity and affirmative action laws and the need for civil rights regulations which address the specific concerns of plant relocation and capital mobility. These findings simply indicate that in and of themselves affirmative action programs offer limited promise. A truly effective black liberation struggle depends on the fusion of efforts to achieve what are currently viewed as two distinct objectives—economic democracy and civil rights.

Notes

1. Vernon M. Briggs, Jr., "The Negro in American Industry: A Review of Seven Studies," *Journal of Human Resources* (Summer 1970), cited in Harold M. Baron, *The Demand for Black Labor* (Cambridge: Radical American, 1971), p. 38.
2. Ibid.
3. William J. Wilson, *The Declining Significance of Race* (Chicago: University of Chicago Press, 1978); Charles Vert Willie, ed., *Caste and Class Controversy* (Bayside, N.Y.: General Hall, Inc., 1979).
4. William K. Tabb, *The Political Economy of the Black Ghetto* (New York: W.W. Norton & Co., 1970); Victor Perlo, *Economics of Racism USA* (New York: International Publishers, Inc., 1975); James and Grace Boggs, *Racism and the Class Struggle* (New York: Monthly Review Press, 1970); James A. Geschwender, *Class, Race, and Worker Insurgency* (New York: Cambridge University Press, 1977).
5. Jeremy Rifkin and Randy Barber, *The North Will Rise Again* (Boston: Beacon Press, 1978); *Plant Closings* (Washington, D.C.: Conference on State and Local Policies and the Ohio Public Interest Campaign, 1979); Bennett Harrison and Barry Bluestone, *Capital and Communities: The Causes and Consequences of Private Disinvestment* (Washington, D.C.: The Progressive Alliance, 1980).
6. Carol L. Jusenius and Larry C. Ledebur, *Documenting the "Decline" of the North* (Washington, D.C.: Economic Development Administration, U.S. Department of Commerce, 1978), pp. 2, 3, 11; David Birch, "The Job Generation Process," MIT Program on Neighborhood and Regional Change, 1979), p. 4.
7. Harrison and Bluestone, *Capital and Communities*; Birch, "Job Generation Process."
8. U.S. Department of Commerce, Bureau of the Census, "Social and Economic Characteristics of the Metropolitan and Nonmetropolitan Population: 1977 and 1970," *Current Population Re-*

ports, Special Studies Series P-23, No. 75 (1978); U.S. Department of Commerce, Bureau of the Census, "Social and Economic Characteristics of the Population in Metropolitan and Nonmetropolitan Areas: 1970 and 1960," *Current Population Reports*, Series P-23, No. 37 (1971).

9. Ibid.

10. Marianne C. Nealon, memo to Richard L. Thomas, President, The First National Bank of Chicago, Sept. 21, 1977.

11. Data supplied by Illinois Commerce Department, reported in "City Short on Economic Zip," *Chicago Sun-Times*, Jan. 23, 1980.

12. U.S. Department of Commerce, Bureau of the Census, "Demographic, Social, and Economic Profile of States: Spring 1976," *Current Population Reports*, Series P-20, No. 334 (1979); U.S. Department of Commerce, Bureau of the Census, *U.S. Census of Population: 1960*, vol. I, pt. 1 (1964).

13. *U.S. Census of Population: 1960*, vol. 1, pt. 1. U.S. Department of Labor, "Work Experience and Earnings in 1975 by State and Area," Report 536, 1978.

14. Rifkin and Barber, *North Will Rise Again*, pp. 30–33.

15. U.S. Department of Commerce, "Demographic, Social, and Economic Profile of States: Spring 1976"; *U.S. Census of Population: 1960*, vol. 1, pt. 1.

16. Shelley Amdur, Samuel Friedman, and Rebecca Staiger, "Investment and Employment Tax Credits: An Assessment of Geographically Sensitive Alternatives" (Washington, D.C.: Northeast-Midwest Institute, 1978), pp. 1, 2.

17. Robert M. Frank and Richard T. Freeman, *Multinational Corporations and Domestic Employment* (Ithaca: Cornell University, 1976), cited in Edward Kelly, *Industrial Exodus* (Washington, D.C.: Conference on Alternative State and Local Public Policies, 1977), p. 2.

18. Ibid.

19. Peggy Musgrave, *Direct Investment Abroad and the Multinationals: Effects on the U.S. Economy* (Washington, D.C.: U.S.

Government Printing Office, 1975), cited in Edward Kelly, *Industrial Exodus* (Washington, D.C.: Conference on Alternative State and Local Public Policies, 1977), p. 7.

20. "Broadening the Ownership of New Capital: ESOP's and Other Alternatives," study prepared for the Joint Economic Committee of the Congress of the United States, 1976, p. 1.

21. Ibid., p. 14.

22. Federal Trade Commission, *Economic Report on Corporate Mergers* (Washington, D.C.: U.S. Government Printing Office, 1969), p. 5.

23. Peter Bearse, "Plant Closings," unpublished manuscript, March 10, 1978; "Indicators for Measuring the Community Costs of Plant Closings," report prepared for the Federal Trade Commission by C & R Associates, Nov. 1978. David A. Smith, with Patrick J. McGuigan, "Youngstown Is Not Unique: The Public Policy Implications of Plant Closings and Runaways," paper prepared for the National Center for Economic Alternatives (Sept. 1978).

24. Marc Stepp (of United Auto Workers), testimony on the National Employment Priorities Act before the Subcommittee on Labor Standards of the House Committee on Education and Labor, August 15, 1978, p. 4.

25. C & R Associates, *Measuring the Community Costs*, pp. 4, 10, 27; Stephen S. Mick, "Social and Personal Costs of Plant Shutdowns," *Industrial Relations* (May 1975), p. 205; Walter G. Strange, "Job Loss: A Psychosocial Study of Worker Reactions to a Plant-Closing in a Company Town in Southern Appalachia," unpublished and undated abstract for a study published by the National Technical Information Service in Springfield, Va., in 1977.

26. Sidney Cobb and Stanislav V. Kasl, *Termination: The Consequences of Job Loss* (National Institute for Occupational Safety and Health, U.S. Department of Health, Education, and Welfare, 1977), p. 179.

27. Roger W. Schmenner, "The Manufacturing Location Decision: Evidence from Cincinnati and New England" (Harvard-MIT Joint Center for Urban Studies, 1978), p. 8.

28. "Results of IMA Survey of Chicago Manufacturer's Problems—Principal Woes Point to Springfield," *IMA Executive Memo* (weekly publication of the Illinois Manufacturer's Association), Jan. 22, 1980.

29. *Akron Beacon Journal*, Feb. 20, 1977, cited in Edward Kelly, *Industrial Exodus* (Washington, D.C.: Conference on Alternative State and Local Public Policies, 1977), p. 3.

30. "Unions in the Sunbelt," *Business Week* (May 17, 1976), cited in Rifkin and Barber, *North Will Rise Again*, p. 33.

31. Ibid.

32. Kelly, *Industrial Exodus*, p. 3.

33. Federal Trade Commission, *Economic Report on Corporate Mergers*, pp. 142–59.

34. Mark Nadel, *Corporations and Public Accountability* (Lexington, Mass.: D. C. Heath and Co., 1976), p. 122; Ritchie P. Lowry, "A Sociological View of Corporate Mergers," paper presented at the 29th Annual Meeting of the Society for the Study of Social Problems, Boston, August 27, 1979; Bennett Harrison, "Testimony in Support of the Notification and Assistance Act (S-127)" before the Joint Committee on Commerce and Labor, Massachusetts General Court, p. 4; William Foote Whyte, "In Support of the Voluntary Employee Ownership and Community Stabilization Act," March 20, 1978.

35. David Moberg, "Shuttered Factories, Shattered Communities," *In These Times*, June 27–July 3, 1979; Gar Alperovitz and Jeff Faux, "The Economics of Urban Reconstruction," *Church & Society* (July–Aug. 1978), pp. 26–34.

36. William Foote Whyte to Isabel Sawhill, Director, National Commission for Employment Policy, July 17, 1979.

37. Jack E. Nelson, "The Impact of Corporate Suburban Relocations on Minority Employment Opportunities," a research paper prepared under contract with the Equal Employment Opportunity Commission, 1974, pp. 30, 32, 45.

38. Birch, "Job Generation Process," p. 22.

39. *The State of the Region: Economic Trends of the 1970s in the*

Northeast and Midwest (Washington, D.C.: Northeast-Midwest Institute and the Northeast-Midwest Coalition, 1979), pp. 29, 30, 69, 70.

40. U.S. Commission on Civil Rights, *Equal Opportunity in Suburbia*, 1974, p. 44.

41. Rifkin and Barber, *North Will Rise Again*, p. 60; "Two Urban Initiatives: A Report Card" (Washington, D.C.: Northeast-Midwest Institute, 1979), p. 4.

42. *Economic Dislocation: Plant Closings, Relocations and Plant Conversion*, Joint Report of Labor Union Study Tour Participants (1979), p. 7; Stepp, testimony before the House Subcommittee on Labor Standards, p. 5.; Smith and McGuigan, "Youngstown Is Not Unique," pp. 8, 9; Don Stillman, "The Devastating Impact of Plant Relocations," *Working Papers* (July-Aug. 1978), p. 47.

43. "Right to Work" pamphlet distributed by AFL-CIO, cited in Rifkin and Barber, *North Will Rise Again*, p. 36.

44. Bennett Harrison and Sandra Kanter, "The Great State Robbery," *Working Papers* (Spring 1976), p. 62; Schmenner, "The Manufacturing Location Decision," p. 11; Stillman, "Devastating Impact of Plant Relocations," p. 47; Kelly, *Industrial Exodus*, pp. 22, 23; Birch, "Job Generation Process," p. 8; Ralph Nader and Jerry Jacobs, "Battle to Lure Industry Costly," *Chicago Tribune*, Nov. 12, 1979.

45. U.S. Department of Commerce, U.S. Bureau of the Census, "The Social and Economic Status of the Black Population in the United States: An Historical View, 1790–1978," *Current Population Reports*, Special Studies Series P-23, No. 80 (1979), p. 15.

46. Ibid., pp. 13, 36.

47. "A New Strategy for Minority Business Enterprise Development" (Executive Summary), James H. Lowry and Associates, Chicago, Illinois, p. 2 and exhibit 5; James A. Hefner, "The Economics of the Black Family From Four Perspectives," in Charles Vert Willie, ed., *Caste and Class Controversy* (Bayside, N.Y.: General Hall, Inc., 1979), p. 83.

48. U.S. Bureau of the Census, "Social and Economic Status of the Black Population," p. 218.

49. Thomas M. Rohan, "Requiem for a Factory," *Industry Week*, Jan. 31, 1977, p. 40; Equal Employment Opportunity Commission Memorandum, *Congressional Record—Senate*, Feb. 22, 1972, pp. 4925–4927; *Abney* v. *Budd*, "Status Conference Report," U.S. District Court, Eastern District of Michigan, Southern Division, Feb. 27, 1979; Suburban Action Institute, Petition to the United States Department of Labor for a Ruling That Union Carbide is in Violation of Federal Contract Compliance Requirements Pursuant to Executive Order 11246, 1977, p. 5.

50. "Industrial Exodus Hits Minority Workers the Hardest," Ohio Public Interest Campaign, Cincinnati (undated) p. 5.

51. U.S. Bureau of the Census, "Social and Economic Status of the Black Population," p. 218.

52. Franklin D. Wilson, *Residential Consumption, Economic Opportunity, and Race* (New York: Academic Press, 1979), p. 147.

53. Nelson, "Impact of Corporate Relocations," pp. 32, 33.

54. John Shields (former manager of Automobile Club of Michigan), affidavit of March 27, 1974, EEOC files.

55. U.S. Commission on Civil Rights, *Understanding Fair Housing* (1973), p. 5.

56. Stanley Masters, *Black-White Income Differentials: Empirical Studies and Policy Implications* (New York: Academic Press, 1975); Wilson, *Residential Consumption*.

57. John F. Kain, *Housing Markets and Racial Discrimination: A Microeconomic Analysis* (New York: Columbia University Press, 1975); *The Impact of Housing Patterns on Job Opportunities* (New York: National Committee Against Discrimination in Housing, 1968); Nelson, "Impact of Corporate Relocations"; Charles Melvin Christian, "The Impact of Industrial Relocations from the Black Community of Chicago Upon Job Opportunities and Residential Mobility of the Central City Workforce," Ph.D. dissertation, University of Illinois at Champaign-Urbana, 1975; John F. Kain, "Report on the Impact of AAA's Relocation on Black Employment," testimony presented in the case of *Bell* v. *Automobile Club of Michigan*, U.S. District Court, Eastern District of Michigan, Southern Division, Civil Action No. 39309.

58. The EEO-1 report breaks down a company's workforce by occupational group and by race and sex within each occupational group. The report must be filed annually by all companies with one hundred or more employees and those holding government contracts worth $50,000 and employing fifty or more workers.

59. Joachim Neckare, Chief, Survey Branch, Equal Employment Opportunity Commission, telephone interview, Dec. 26, 1979.

60. Harrison, testimony, before the Joint Committee on Commerce and Labor, p. 6.

61. Christian, Impact of Industrial Relocations, pp. 197–202.

62. Wilson, *Residential Consumption*, p. 194.

63. U.S. Department of Labor, Bureau of Labor Statistics, "Employment in Perspective: Minority Workers," Report 584 (Feb. 1980), p. 1.

64. *Labor Force Information for Affirmative Action Programs 1979,* vol. 1 (Illinois Bureau of Employment Security, 1979).

65. *Bernetha Abney* v. *The Budd Company*, No. 6-71845 (U.S. District Court, Eastern District of Michigan, Sept. 7, 1976); *Bell* v. *Automobile Club of Michigan*.

66. Alfred W. Blumrosen, "The Duty to Plan for Fair Employment: Plant Location in White Suburbia," *Rutgers Law Review* (Spring 1971); Equal Employment Opportunity Commission Memorandum, U.S. Congress, Senate, *Congressional Record*, Feb. 22, 1972, pp. 4925, 4926. (This memorandum created sufficient controversy that EEOC Chairman Willian H. Brown III later issued a statement disavowing this position as EEOC policy. See Samuel C. Jackson and Michael E. Abramowitz, "Housing Transportation and Fair Employment," paper presented at symposium in observation of the tenth anniversary of the EEOC, Rutgers Law School, 1975, p. 3.)

67. 42 U.S.C. § 703 (a)(1)(2).

68. Equal Employment Opportunity Commission Memorandum, p. 4925.

69. *Griggs* v. *Duke Power Company*, 401 U.S. 424 (1971).

70. Ibid., p. 430.

71. Blumrosen, "The Duty to Plan," p. 392.
72. Ibid., p. 395.
73. Suburban Action Institute, "Petition to the United States Department of Labor for a Ruling That Union Carbide Is in Violation of Federal Contract Compliance Requirements Pursuant to Executive Order 11246," 1977; Equal Employment Opportunity Commission, *Affirmative Action and Equal Employment: A Guidebook for Employers, 1974*, vol. 1, pp. 13, 14; vol. 2, app. D; Blumrosen, "The Duty to Plan," pp. 401–4.
74. Executive Order 11246 (1965), pt. II, subpt. 13, sec. 202, p. 1.
75. 41 C.F.R. § 60–2.23 (1971).
76. *Abney* v. *Budd*.
77. *Bell* v. *Automobile Club of Michigan*.
78. Westchester Residential Opportunities, Inc., *Equal Opportunity in Housing: A Manual for Corporate Employers* (Washington, D.C.: U.S. Department of Housing and Urban Development, 1973).
79. Blumrosen, "The Duty to Plan," pp. 388–404; Suburban Action Institute, 1977 "Petition," pp. 40–43; Alfred W. Blumrosen and James H. Blair, *Enforcing Equality in Housing and Employment Through State Civil Rights Laws* (Newark, N.J.: Adminstrative Process Project of Rutgers Law School, 1972), pp. 499–502.
80. Richard B. McKenzie, *Restrictions on Business Mobility: A Study in Political Rhetoric and Economic Reality* (Washington, D.C.: American Enterprise Institute, 1979); U.S. Congress, Senate, *Congressional Record*, 1972, pp. 4927, 4928.
81. William Kapp, *The Social Costs of Private Enterprise* (New York: Schocken Books, 1975); David Smith and Patrick McGuigan, *Towards a Public Balance Sheet* (Washington, D.C.: National Center for Economic Alternatives, 1979); Peter J. Bearse, "Influencing Capital Flows for Urban Economic Development: Incentives or Institution Building?" (Department of Economics and Finance, Baruch College, The City University of New York, 1978).
82. "Measuring the Community Costs of Plant Closings: Over-

view of Methods and Data Sources," report prepared by C & R Associates for the Federal Trade Commission, p. 70.

83. Bennett Harrison and Barry Bluestone, "Capital Mobility and Economic Dislocation," outline of a paper commissioned by the Progressive Alliance, 1979, p. 1.

84. Smith and McGuigan, "Youngstown Is Not Unique"; Bearse, "Influencing Capital Flows"; Stewart E. Perry, "Worker and Community Ownership: Urban Jobs and the CDC," research proposal for study of community development corporations; "Plant Closing Legislation and Regulation in the United States and Western Europe: A Survey," report prepared by C & R Associates for the Federal Trade Commission, 1979.

85. Bearse, "Influencing Capital Flows," p. 10; Perry, "Worker and Community Ownership," p. 14.

86. Bearse, "Influencing Capital Flows," p. 9–11; Perry, "Worker and Community Ownership," p. 14.

87. C & R Associates, "Plant Closing Legislation and Regulation," pp. 27–46.

88. Ana Gutierrez Johnson and William Foote Whyte, "The Mondragon System of Worker Production Cooperatives," *Industrial and Labor Relations Review* (Oct. 1977).

89. William Schweke, *Plant Closing Strategy Packet* (Washington, D.C.: The Progressive Alliance and the Conference on Alternative State and Local Policies, 1980).

90. P.A. 44, Michigan, 1979.

91. "Fraser to Be on Chrysler Board," *Chicago Tribune*, Oct. 26, 1979.

92. Ibid.

93. Ron Shinn, "Workers Had No Warning of Firestone Plant Closings," *In These Times*, April 9–15, 1980.

94. "The Role of the Federal Government and Employee Ownership of Business," Select Committee on Small Business, United States Senate, Jan. 29, 1979, p. iii.

95. Ibid.

96. Daniel Zwerdling, *Democracy at Work* (Washington, D.C.: Democracy at Work, 1978); *Employee Ownership*, Report to the

Economic Development Administration, U.S. Department of Commerce (Ann Arbor: University of Michigan, 1977); Karl Frieden, *Workplace Democracy and Productivity* (Washington, D.C.: National Center for Economic Alternatives, 1980); Jaroslav Vanek, ed., *Self-Management: Economic Liberation of Man* (Baltimore: Penguin Books, 1975), pp. 11–36; Charles Hampden-Turner, "The Factory as an Oppressive and Non-Emancipatory Environment," in Gerry Hunnious, G. David Garson, and John Case, eds., *Workers' Control: A Reader on Labor and Social Change* (New York: Vintage Books, 1973), pp. 30–45.

Chapter 4

Gaining Control Over Our Economic Resources

Randy Barber

The basic function of any economy is to organize and allocate the resources of a society, be it poor or wealthy, agrarian or industrial. In theory, an economy should function, first, to match the most pressing needs of a people with the resources at hand, and, secondly, to expand their potential to produce and consume.

In the past, a major function of government has been to identify needs that the private economy was not fulfilling and to develop programs that would. Clearly, the intent of many of our current elected leaders is to reverse this trend and to give the "free markets" an even greater role in determining which goods and services are made available to whom. The belief—or hope—is that these markets will ultimately provide more and better goods and services than "wasteful and inefficient" government programs ever would, even in areas of clear market failures.

The impact of this move is likely to exacerbate a number of major problems that government programs have attempted, however feebly, to address. First, capital markets are highly liquid and mobile, operating essentially on a global scale. The new wave of tax cuts and accelerated depreciation will probably *increase* the mobility of capital, making it even easier for corporations to relinquish current investments in favor of more lucrative opportunities—often in other regions, countries and economic sectors. For example, U.S. Steel will probably invest more money in the banking, petrochemical, agricultural, and other nonsteel activities that it has developed, at the expense of its investments in steel.

The impact that this increased mobility of capital will have on labor, low income people, and distressed communities and regions is likely to be enormous. The balance of power in our workplaces and our communities will be shifted even further in favor of those who control the allocation of capital. This allocation process has become increasingly concentrated over the past decade, and it is reasonable to assume that the Reagan administration programs will accelerate it even more. The fact that corporate profits will be increased is, in many ways, a much less important issue than the fact that corporations will be able to shift resources under their control even more easily. The administration and congress have decided to let the private financial markets be the final arbiter of which needs will be met and which will go unmet. Accordingly, the priorities for resource allocation will be, more than in the past, established in the corporate boardrooms. Thus the public sector will be even less effective at counteracting the disruptions created by private economic decisions.

While we surely shouldn't just meekly give up hard won gains and well-reasoned principles regarding the desired role of the public sector in the economy, we will, nonetheless, need to develop effective strategies that respond to this new reality. One of the major tasks will be to identify sources of capital that can help fill the gaps left by the abandonment of an activist public role in the economy. This effort will require what can best be termed "economic organizing": matching capital resources with people and their needs.

As we look at the world today, one thing should be very clear: we are entering a new economic era. These are times like none we have ever experienced, not only in this country but also around the world. There is a growing relative scarcity of basic resources: water, minerals, land, energy and so forth. We are entering an era of competition for *access* to increasingly sought-after resources. There are cartels forming in many basic resources, and our economies have a range of problems today that they have never before confronted. In the future, it seems obvious that our economy will not repeat the economic boom-times of the forties, fifties and sixties, periods in which many of our broadest social programs and many of our jobs were created. There will be growth in the economy, but it is likely to be much more selective and skewed than in the past.

By now, the terms "reindustrialization," "economic revitalization" and Ronald Reagan's "making America great again" have become over-used to the point of seeming meaningless. Nonetheless, the underlying meaning of these phrases is of critical importance to our communities and to our workers. Beneath the rhetoric, the fact is that the American economy is being re-made in many significant ways. The auto industry has suffered what will likely be a permanent reduction of one-third of its capacity—and jobs. Other basic manufacturing industries are suffering a similar fate. On the other hand, the massive concessions which the Carter and Reagan administrations have made to corporate America not only presage a shift in the balance in power in our society, but also create a new dynamic for economic change and growth.

Implicit in the now-established tilt towards corporate profitability over all other considerations is a central issue of the 1980s: there is a new equation in the *politics of priorities.* The process of setting societal priorities inevitably forces a decision about who is more important and who is less important; whose needs will be met and favored, and whose needs will go unmet. The economic advisors, not only of Ronald Reagan, but also of the two other major presidential candidates as well, all said that we must consume less; we all have to sacrifice; we have to save and invest more to increase our productive capacity.

The crucial question in the politics of priorities is this: Who will have to sacrifice, how will that be determined, and who will benefit from the sacrifices? The decontrol of oil and natural gas is the kind of priority decision which represents one way of deciding how sacrifices will be spread throughout the population. It represents a concrete strategy for the allocation of resources within our economy.

With the heightened competition for access to the resources of our society, the even more crucial question becomes: How are those resources allocated? Today, the resources of our society are allocated primarily through the investment process and through what are known as the capital markets. Investment capital is purely and simply *money.* The investment of capital is the key to the creation of jobs, the building of factories and the purchase of plant and equipment. The withholding of capital has precisely the reverse effect: jobs are destroyed, plants are closed and equipment wears out.

This is the root of an incredible irony, and opportunity as well: The capital that will reindustrialize, or rebuild, America, will come, in large part, from the deferred wages of fifty million American workers, through their pension funds. Pension funds are the largest single source of capital in our economy. They are worth almost $700 billion, and they're growing at a rate of 10 percent per year. In 1982, pension funds will invest between $70 billion and $80 billion new dollars in the economy, one out of four new dollars that will be invested in the entire economy. Pension funds own 20–25 percent of corporate stock and 40 percent of corporate bonds. It is projected that, by 1995, pension funds in the private and public sector will be worth almost $4 *trillion* and will own over one-half of the stocks and bonds of American companies (For more complete data on pensions, see Tables 4-2, 4-3, and 4-4 in the appendix to this chapter.)

Pension funds are *the major source* of individual savings in the economy today. Having displaced local savings institutions in that role, this form of savings has moved workers' money out of localized capital markets into the national and global capital markets. This means that savings which were once principally dedicated—by law—to investments in local housing and small businesses are now both far more mobile and invested in a far more concentrated and centralized market.

Unions have been a major factor in the growth of pension funds. Although unions represent only 22 percent or 23 percent of the private sector workforce, two-thirds of all participants in private sector pension plans are in collectively bargained plans, because pension funds are a basic "fringe benefit," something that unions have been pursuing since the 1940's. Pension funds are legally defined as "deferred wages," they are not the gift of the employer, and they are not the property of the employer when placed in a pension trust. There are four major types of pension funds: federal government funds, state and local government funds, corporate controlled funds, and jointly controlled ("union" or multiemployer funds):

- Federal retirement funds, with about $120 billion in assets, are invested exclusively in federal treasury issues. These funds are controlled by government administrators. Since they are not invested in the private capital markets, these funds will not be mentioned further in this paper.
- State and local government public employee pension funds have some $200 billion in assets, which are invested primarily in corporate bonds and stocks, and to a lesser extent in federal treasury issues, real estate, mortgages and other assets. The control of these funds varies from state to state, but often includes some representation from public employees and their organizations, as well as public officials and members of the financial sector. State and local legislative bodies set the broad investment policies of these funds.
- In the private sector, corporate—or single employer—pension funds, worth some $330 billion, are invested in a range of stocks, bonds, mortgages, real estate, federal issues and other securities and investments. As the name implies, these funds are controlled exclusively by corporate management, although roughly half of their assets are subject to collective bargaining by unions.
- The jointly controlled, multiemployer or "union" pension funds are worth about $50 billion (although estimates range from $35 billion to $92 billion since no comprehensive survey has ever been conducted). These funds invest in much the same way as corporate funds, except that there is a somewhat greater bias in favor of fixed income investments like bonds and mortgages. Joint funds are concentrated in highly competitive industries with many small employers and relatively great turnover of workers: the construction industry, transportation, textiles, garment manufacturing, retail services, health care, food processing, mining, and so forth. The Taft-Hartley Act prohibits unions

from exercising any more than 50 percent control over a pension or other trust by requiring that management be equally represented. It should be noted that there are no requirements that workers be represented on the boards of corporate funds.

Typically, a pension fund turns its assets over to an outside investment manager, although there is a marked trend to "in-house" management of some large corporate and public employee funds. The pension investment "industry" (with about $3 billion in "sales") is quite concentrated with just a few hundred firms of any significance at all. The top 10 money managers controlled $120 billion in pension assets at year–end 1979 and the twenty-five largest controlled $200 billion. In addition, some $141 billion in pension assets were managed by "in-house" investment professionals.

The Pension Investment Process

The 1970's were a quite turbulent time for the investment industry. At the beginning of the decade, there was the so-called "go-go" period when stock prices rose dramatically. This was followed by a major downturn in 1973 and 1974 (pension funds alone lost an estimated $21 billion in 1974 and some funds suffered net declines in asset values of 30–40 percent). During the rest of the decade, stock prices were generally unchanged while interest rates fluctuated wildly. According to one major study, the average *corporate* pension fund experienced an annual return of 3.9 percent on its stock investments for the ten-year period 1970–1979, while the stock market as a whole (as measured by the Standard & Poors 500, a broad composite of the stock market) had an average annual return of 5.9 percent for the same period. This same study (conducted by A. G. Becker's Fund Evaluation Service) found that for the fifteen-year period 1965–1979, corporate pension funds had a total annual return on all of their investments that averaged 4.1 percent, while inflation for that period averaged 6.2 percent per year as measured by the Consumer Price Index. In 1980 and early 1981, the stock market rebounded significantly while bond prices fell as interest rates and inflation reached record high levels.

The poor relative performance of "active" professional investment managers has led to the creation of a new investment vehicle known as the "index fund." This highly computerized mechanism simply attempts to mirror the stock market as a whole. Since it has been shown that, in terms of investment performance, throwing darts at the stock quotations is as effective as spending millions of dollars for in-depth research, pension funds are increasingly opting for these much less expensive vehicles which are designed to perform as well, or as poorly, as the market. In the past five years, almost $10 billion in pension assets have been placed in index funds.

The turmoil in the financial markets caused pension portfolio managers to vary their investment strategies dramatically during the past decade, first moving heavily into stocks, then into bonds, and finally back into stocks again, as well as into a whole range of "alternative" investments such as real estate, gold, diamonds, antiques, commodities, financial futures (speculation on future interest rates), direct international investments and a host of others.

Although there are wide and ever-changing variations, private sector pension funds, on the average, have about 50 percent of their assets invested in corporate stocks, 30 percent in corporate bonds, 10 percent in short term "cash equivalents" and 10 percent in other investments such as real estate, mortgages, and long-term government bonds. State and local public employee pension funds reverse the ratio of equity and debt, having, on the average, about 30 percent invested in stocks, and 50 percent in bonds.

Pension fund investments are highly concentrated in the securities (stocks and bonds) of the largest American corporations. This preference follows a broader pattern of what has been called the "institutionalization" of the capital markets, where individual investors have progressively been replaced by institutions such as insurance companies, banks, mutual funds and pension funds. Institutions account for about 45 percent of all stock ownership and over 80 percent of bond holdings, and pension funds represent well over half of all institutional assets. In terms of the stock market, these institutions are far more important than even their proportion of ownership would indicate: they account for over 75 percent of all trading on the various stock exchanges in dollar volume (see Table 4-1 and Figure 4-1 in the appendix to this chapter). The stocks of the largest corporations, known as the "blue chip" issues or the "nifty fifty" in Wall Street jargon, are easily bought and sold in very large volumes. In contrast, institutional investors largely avoid the securities of companies with under $100 million or so in assets. This is partially explained by the fact that larger blocks of the big companies' securities can be traded without affecting the price of the security (in financial terms, they are more "liquid").

This tendency leads to a second problem: if smaller blocks are to be traded, more separate investment decisions must be made to invest the same amount of money. A further problem with investing in smaller firms is that there is less generally available information on them, making it more time-consuming and expensive to make an investment. This vicious cycle has led to the creation of a "multi-tiered" capital market, where the largest corporations have no difficulty in raising capital, but small and even rather large firms have great problems securing needed funds, especially during periods of "tight" money, high interest rates, recession and general economic uncertainty.

As previously noted, pension investors had a substantially lower return on their stock investments than did the stock market as a whole during the last decade. This difference is partially explained by the fact that the stocks of smaller companies had the best performance, while larger companies were less profitable than the average. In fact there was a distinctly reverse relationship between the size of a firm and the profitability of investments in its stock. Since pension fund investments were concentrated in the equities of the largest companies, it should not be surprising they underperformed the broad market averages.

There is a second, related factor in pension funds' below average performance. Because the management of these assets is also concentrated, pension fund and other institutional investors compete against each other. The net result is lower average returns for all. This phenomenon has been formalized into what is known as the "random walk theory," derived from academic studies of market efficiency and pricing. This theory holds that future price changes in the market value of particular stocks are random; therefore it is impossible to predict stock performance based on available current knowledge. Ironically, the random walk theory works precisely because the market is dominated by institutions. Since they all have huge research budgets, there is little relevant information, especially on the largest companies, that is not known by all institutional investors. The evidence for this theory has been collected over the past decade and has come to be largely accepted by academics and a growing number of professional investment managers.

This has translated into below average performance for pension funds. For instance, between 1962 and 1975, 87 percent of all investment managers underperformed the Standard & Poors 500 averages. Moreover, a study comparing the performance of large bank trust departments with those of much smaller bank trust departments, which typically spent little on research and trade less actively, found that the large and "sophisticated" money managers underperformed the small managers.

Pension funds, almost by definition, are long term investors. Their goal is to pay benefits that can be anticipated as much as sixty years or more in the future. While pension managers and administrators have been criticized for behaving as though these funds have a very short-term horizon, most will acknowledge that their basic task is to secure long-term income and appreciation.

The typical pension fund—public or private—takes in much more in contributions and investment income than it pays out in benefits and expenses. This disparity means that, in theory at least, only a relatively small portion of pension assets needs to be invested in highly "liquid" securities to be sold on very short notice for cash. In practice, most pension investments are quite liquid. On average, 10 percent to 30 percent of a fund's assets are "turned over" in a typical year, as portfolio managers sell less profitable investments for more profitable ones, as short term investments mature, and as funds realize gains from past investments. Although there is no real need for such behavior as it affects an entire portfolio, most pension funds avoid investments that are not readily marketable.

Of special interest to advocates of "alternative" pension investments are two factors that affect the cost of investing, and thus the cost and availability of pension capital. These factors are called information and transaction costs. Both of these factors can lower or raise the expected return on an investment. Information and transaction costs are basic considerations for any investment. Information is valued because the more an investor knows about a particular investment, the easier it is to make a decision and evaluate the potential risk and return. One of the reasons for the great concentration of pension investments in the securities of the largest corporations is that there is a tremendous amount of easily obtained information on them. The more difficult it is to obtain information on a potential investment, the greater the perceived risk and the higher the expected return—this in addition to the need for a greater return to compensate for the increased expense and effort of evaluating such an investment in the first place.

Transaction costs are incurred whenever a security is bought or sold. Included in these costs are brokerage commissions, transfer and registration fees, research and information costs, and administrative expenses to the fund. A basic consideration here is the ratio of transaction costs to the price of the purchase or sale. This is another source of built-in bias in favor of the securities of the largest corporations. These securities are easily bought and sold in very large blocks, and these transactions take little time to complete. Brokerage commissions are negotiated so that they decline as a percentage of the value of a transaction as its size increases. Thus, it is not only much easier but also much less expensive to buy or sell one $10 million block of AT&T than it is to buy or sell twenty $500,000 blocks of the stock of much smaller companies, even if they are just as marketable as the AT&T stock. It also takes a lot more time to decide on twenty different investments than it does on one. Thus, those stocks could even be somewhat more profitable than AT&T, but the fund would still buy the Ma Bell securities.

The Criteria for Pension Investments

The broader question of the appropriate criteria that should be used in determining pension fund investment policies is obviously crucial. For the most part, control of these funds is effectively in the hands of investment professionals, even where there is union or public sector representation. These professionals follow standard institutional financial criteria, and rarely attempt to differentiate between various sources of capital and its ultimate use. There is even some philosophy behind this practice, that of capital market "efficiency:" capital markets respond to changing conditions by flowing to the most productive investments and away from unproductive ones. This perspective was perhaps best articulated by James LaFleur, former director of the $5 billion Wisconsin Investment Board, the public employee pension fund system for that state:

> [C]apital is extremely mobile and flows to where it's wanted and rewarded. It is by far the most mobile of the elements of production which also include land, labor and management. If it's punished by taxation, regulation or other means, it flees. It doesn't vote and it doesn't complain. It just moves to the highest bidder and always seems to be in short supply to someone. It is an elegantly simple process with flows dictated by a rather efficient market system. . . . Intranational and even international borders are not significant barriers. [Speech to Municipal Finance Officers' Association, Detroit, Michigan, June 1979].

Simply, LaFleur is saying—and other investment professionals would certainly agree—that anything truly worth doing will be done and any real need will be met if "capital" is allowed to do its job unfettered. He holds that the capital markets are, and should be, the autonomous and self-regulating allocators of societal resources. The criteria for this process are determined by what is in the best interest of capital, as defined by those who are its intermediaries in the marketplace. And the capital markets are presented as independent and neutral enforcers of economic efficiency and of the supremacy of financial considerations over social or political concerns. In fact, social good is derived from, and even defined by, the efficient functioning of these markets.

Needless to say, not all would agree with these standards for the investment of pension, or other, capital. It should be noted that a process such as LaFleur describes fundamentally discriminates against workers who have been able to achieve decent wages and working conditions. It also discriminates against countries and communities which have been successful in efforts to force companies to absorb at least part of the "external" costs (pollution, industrial diseases, economic dislocation, etc.) created by their activities. This situation also implies that every time workers are able to gain increases in pension benefits, the funds deposited to pay for those increases will be invested under criteria viewing those workers' jobs as less desirable investments precisely because of the greater cost that the employer must now bear. It is, one could say, an ironic economic "catch-22."

Of course, the humor of it all is lost on workers who are without jobs or in communities which suffer from capital disinvestment. And this is precisely where unionists, many public officials, and activists have first become interested in the issue of pension fund investment and control. To them the issue increasingly can be framed as follows: In an era of accelerating competition for *access* to relatively scarce capital, every effort should be made to establish a link between the source of capital and its ultimate use, and capital such as pension funds should be "recycled" in ways that have a beneficial, direct impact on pension fund owners through job creation and community stabilization. At the very least, pension funds should not be invested in ways that work counter to the interest of pension fund participants.

Writing in the introduction to a report (to be discussed below) of the AFL-CIO Executive Council last August, John H. Lyons, president of the Ironworkers Union and vice-president of the AFL-CIO, notes that the report finds "pension funds are invested in companies which are among the most anti-union, export workers' jobs to low-wage countries, ignore workers' needs for health and safety protection and in other ways hinder rather than help workers in the achievement of their most basic and legitimate objectives." Further, says AFL-CIO president Lane Kirkland, "The primary purpose of pension funds is, with prudent investments, to secure retirement benefits for workers. Without jeopardizing this primary objective, pension funds should invest to improve the life and conditions of working men and women. Unions should be in a position to influence the investment of pension funds to better serve the interests of workers."

Developing Alternative Investment Criteria and Mechanisms

It is one thing to decide that pension funds are being invested in the wrong ways, but it is an entirely different task to determine what should be done with them. Neither unions nor the public sector have much experience with broad investment questions or with specific investment vehicles. Thus, the proposals that are advanced for "alternative" pension fund investment programs tend to be either quite sweeping with few details or reduced to suggestions for very specific and limited new investment mechanisms. Both tendencies are useful, having helped clarify many issues, but it should be stressed that there remain large areas which are relatively undeveloped.

The previously mentioned AFL-CIO Executive Council report represents the first time that this body has thoroughly addressed the question of pension investment and control. The Executive Council, in approving this report, established four policy goals for union participation in pension fund management:

- "To increase employment through reindustrialization, including manufacturing, construction, transportation plus maritime and other sectors necessary to revitalize the economy.
- "To advance social purposes, such as workers housing and health centers.
- "To improve the ability of workers to exercise their rights as shareholders in a coordinated fashion.
- "To exclude from union pension plan investment portfolios companies whose policies are hostile to workers' rights."

A major feature of the AFL-CIO position was the call for the creation of a "Reindustrialization Board" which "should be directed by a tripartite board of directors, equally representing the labor movement, employers and the public." This board would be established by congress with the authority to facilitate the investment of pension funds for reindustrialization purposes through the granting of guarantees for a minimum level of return to pension funds that invest in such a program. This proposal was worked out with the Carter Administration and is now obviously dormant. Ronald Reagan, while a candidate, strongly criticized this notion, but there have been several reports in the business press that at least some of his advisors are shaping similar proposals.

The Executive Council report also contains recommendations about investments in housing, shareholder resolutions, avoidance of blatantly anti-union companies and the need for unions without any voice in the investment of negotiated pension funds to gain a voice.

This last recommendation was actually a reiteration of a report issued by the Industrial Union Department of the AFL-CIO. The IUD report, released in June of 1980, contained a detailed study of the investments of about 200 different pension funds, and it set forth rather detailed proposals for unions without control over pension investments to follow. These suggestions include demands for information, for joint administration, and for specific investments such as housing, and community development.

The Building and Construction Trades Department of the AFL-CIO began a training and education program for pension fund trustees of its affiliates during the fall of 1980. These unions are almost exclusively involved in jointly controlled pension funds, which are worth some $25 billion in that industry. The focus of these programs was quite specific: how can trustees of these funds legally invest in union-built structures that their member-participants put up? There are already dozens of examples of how this investment can be made, and the training program was designed to make other trustees aware of them.

Some of these building trades pension fund investment initiatives include:

- A group of Southern California funds that have joined together to form a "Foundation" which facilitates investments in union built structures, and is currently placing about $15 million a month in such investments.
- About 20 percent of the assets of the Milwaukee building trades combined pension fund that is invested in union built structures, including several moderate income apartment units.
- An Operating Engineers local union in Ft. Lauderdale, Florida that is investing significant portions of its portfolio in construc-

tion and real estate development and is offering mortgage loans to participants at somewhat lower interest rates.

- Several funds in the St. Paul/Minneapolis area that are participating in several programs for mortgage investments, including one that is geared for moderate income households. These funds are also interested in investments in substantially rehabilitated structures.
- The International Brotherhood of Electrical Workers national fund that has over 40 percent of its assets invested in mortgages, much of this in housing related investments.

A number of other international and local union funds have begun their own investment programs, usually concentrating on ways to stabilize employment in their industry. To date these have included only the construction and maritime industries, but there are indications that unions in several other industries are exploring new job-creating investment initiatives as well. The major problem, however, is that new investment mechanisms must be created—or at times, new government programs enacted—to make these programs viable in other sectors of the economy. Some of these needed actions are being developed in the public sector at the state and local level. There have been a large number of studies of the investments of public employee pension funds at the state and local level. Several of these reports have developed proposals for "strategic investment targeting" or "selectively steering pension capital" with the purpose of local economic development and job creation.

The first wave of these studies focused on developing new approaches to utilizing existing investment vehicles or creating new ones where necessary. This included federal investment guarantee programs to target sectors of the economy such as small business, alternative energy production and housing. State and local economic development agencies were usually singled out for a role in this process. By utilizing the resources of these agencies, investments could be realized that would not have otherwise been made. These agencies would absorb the initial information and transaction costs that often present major barriers to desired economic activity and investments.

These proposals represent a broader trend in public involvement in the economy. During the last half of the seventies, a new group of economists emerged who specialize in "development finance." Their goal is to identify ways in which the public sector can encourage new investments and create needed investment vehicles. Dissatisfied with the traditional state and local government approach of luring large new factories through various grants and subsidies, these economists have advocated efforts to facilitate the creation of new firms and the growth of small, young companies. They point to data showing that most new employment and the bulk of economic growth and innovation come from these and not the large, established firms. Venture capital, development banks, targeted tax policies and technical assistance are the main tools that these economists propose. While there are differences among these economists, they share the desire to identify market functions—both positive and negative, from a broader perspective—and then organize those functions in ways that achieve defined public purposes.

This has led to a range of proposals for public pension fund investments in smaller firms, through development banks, and through various "secondary market" vehicles, which purchase investments made by others, pool them, and sell them to the pension fund. By far the most sophisticated proposals have emerged from a task force established by California Governor Jerry Brown. The task force has adopted a wide range of proposals developed by its staff and several committees in the areas of renewable energy, small business, industrial re-investment, economic development and affordable housing. The underlying theme of this task force is the need to create new mechanisms, to establish new investment pools and, generally, to devise programs that will facilitate the prudent investment of pension assets in ways that will bolster the California economy most effectively. There are over $30 billion in public sector pension funds in the state and at least an additional $30 billion in private sector funds as well.

The task force has developed in-depth proposals for a secondary market mechanism for energy conservation and solar energy loans, for the creation of a venture capital fund, for separate debt and equity pools for investment in small but growing firms, for new housing development, and for a state level coordinating agency that would facilitate these programs. The initial goal is to make the investment of about $500 million in these programs over the next two years. Also, the California state legislature appropriated $400,000 in mid-1981 for the creation of a Pension Investment Unit in the governor's office to implement the task force's recommendations. Public employee pension funds have begun to move in new directions in a number of states and cities. Below are some examples:

- In Hawaii, and now California, the public funds make loans to participants for home mortgages. The Hawaii program has over $300 million invested in this manner and offers it somewhat below market loans.
- Several states, including Texas, Massachusetts, Michigan, Connecticut, Alabama and Oregon target their housing investments in-state. The programs of these states alone have placed at least $1 billion in this way over the past three years.
- The cities of St. Paul and Minneapolis have created a fund which pools pension funds with foundation monies and federal funds to offer "shared appreciation" mortgages to families with moderate

incomes. By accepting a share of the increased value in the home instead of interest and principal payments, this program allows families to purchase homes (both new and rehabilitated) that they would not be able to otherwise.

- Proposals have been advanced by public officials and union leaders for pension funds to invest in local economic development programs including energy conservation, small business and industrial development loans.

Public employee unions in Connecticut, New Jersey and Wisconsin have advanced such proposals as part of their collective bargaining demands with the state. Further, governors' task forces have been established in several states besides California.

Legal Considerations in the Investment of Pension Funds

The laws governing the investment of pension funds can be divided into two categories: laws for private sector funds and laws for public sector (state and local government) funds. Since these funds do not pay taxes on their income, however, the federal Internal Revenue Service regulates both kinds of funds to the extent that it has the power to grant and revoke this tax exemption based on a determination of whether the funds are being managed for the purpose of retirement income and not, for instance, as a source of capital for the sponsor.

The basic law governing the investment and administration of private sector pension funds is the Employee Retirement Income Security Act of 1974 (ERISA). The scope of ERISA is quite broad and includes minimum standards for funding pension plans, for vesting of benefits, for administration, record-keeping and disclosure, and for "prudent" investment of plan assets. ERISA establishes standards of conduct for "fiduciaries" (those who have some form of control over the plan's assets), and it provides for legal remedies against fiduciaries who violate these standards. Penalties can include removal from positions of responsibility and personal liability for losses incurred because of imprudent actions. ERISA also establishes a re-insurance system, accomplished through the Pension Benefit Guarantee Corporation, to guarantee at least part of participants' vested benefits should their fund go broke. ERISA does not mandate the establishment of pension plans by an employer, although this has been proposed by a presidential commission; but it does set forth minimum standards for a plan once it is created.

ERISA's provisions covering the investment of plan assets are very general. They require that investments be made in the interest of plan participants and that fiduciaries make investment decisions at "arm's length." The Department of Labor, which is charged with enforcing the law and developing regulations, has taken the position that it will not rule on the prudence of any class of investments in the abstract, but will make determinations only on a case-by-case basis by examining the impact of a particular investment on the overall portfolio of the fund. In theory, any investment is theoretically permissible. In practice, private pension funds have been invested in almost every conceivable type of asset, from corporate securities, to commodities, to real estate, to speculative futures, to mortgages, to gold and diamonds; however, these funds are still concentrated in the stocks and bonds of the largest American corporations.

Except for IRS determinations, state and local government public employee pension funds are regulated by the legislative bodies that have created them. Over the past few years, there has been considerable discussion of extending ERISA to these funds, and this thought has resulted in public pension plans' adopting procedures that are increasingly similar to those required under that law. However, each legislative body is still free to structure investment regulations as it sees fit. For a number of reasons, legislatures have tended to establish specific guidelines for the investment of these funds, often specifying the precise characteristics of permissible investments. The result is what are known as "legal lists." These lists often set forth, in great detail, the investment standards to be met (bond ratings, dividend payment record, total assets of the corporation) and even the percentage of different types of investments that can be held by the fund. If a particular kind of investment is not listed, then it is usually deemed ineligible for investment by the fund. These standards vary widely from state-to-state, but the basic pattern is for higher fixed-income investment levels and lower equity levels than would be found in the typical private sector fund.

Because they tend to be much larger than the average private sector fund, public pension plans are more often administered by "in-house" professionals. Also, to the extent allowed by local legislation, they also tend to become involved in directly negotiating investments with large corporate borrowers for such purposes as plant expansion, the purchase of equipment, the construction of commercial and residential structures and so forth.

The Emerging Issue of Pension Fund Investment and Control

The growth and development of pension funds occurred without any broad debate over how these funds should be invested, or even controlled. Obviously, unions and workers were aware of the fact that pension funds were being created, but their focus was on negotiating the pension *benefit* to supplement Social Security, not on building up a massive pool of investment capital based on the deferred wages or savings of workers. Indeed, unions have been the driving force behind the expansion of the pension system as it exists today in the United States: although unions represent less than one-fourth of the workforce, at least two thirds of all pension funds (in terms of assets) are subject to collective bargaining. Pension plans are a basic "fringe benefit" and they account for between 7 percent and 15 percent of total compensation, depending on the industry and the level of benefits provided.

Beside the fact that pensions were negotiated as a basic benefit and not as an investment pool, there is another major reason for the lack, until recently, of union concern about the investment policies that these funds pursued. American trade unions have historically insisted that they want no involvement whatsoever in "management functions." Clearly, the investment process can be seen as such a function. "If we become actively involved in investing pension funds," one high union official told this author in 1977, "we would certainly be investing in companies that we negotiate with. What if we ended up owning a major interest in one of those companies? We'd end up bargaining with ourselves, and we'd be forced to decide between our role as representatives of workers and as representatives of owners."

Today, that same union official has changed his mind, although he still has some important reservations: "Pension funds are too large and too significant in our economy to leave solely in the hands of companies and money managers. Our members alone have at least $15 billion in the pension funds that we have negotiated, although we have no voice in their investment. Now, that's how we used to want it, but we've realized that we simply can't afford to leave the investment of these funds to others. We've learned what's wrong with the way that these funds are used: they're invested mostly in non-union companies, they're not investing back into our communities where our members live and work, they don't put money into anything but "blue chip" corporations, and they've had a lousy return. As I say, we know what's wrong; but I still worry about what we should be doing. How do we decide what types of companies should get our investments? Should our funds invest in companies that are in trouble, but are organized, or should we invest in companies that are healthy and growing but not yet unionized? And how do we decide between the need for higher wages for workers and more profits for the corporation, and thus for the workers' pension fund? We don't even have experts we can rely on to help us sort through these and other questions. But I guess we have to start somewhere." (Interview, December 1980).

Indeed, the pension fund issue has emerged, over the past few years, as one of the major concerns of unions, public officials and activists on one side, and corporations, money managers and conservative economists on the other side. The underlying debate relates to the control and direction of this massive and rapidly growing pool of investment capital. Weaving through this debate are questions about industrial policy, local economic development, job creation, housing, legal constraints, high interest rates, relative risk of different investments, union-versus-non-union companies, South Africa, market "failures," productivity, economic efficiency, growth and scarcity, capital allocation, new investment mechanisms, social justice, and economic as well as political power. As should be evident, this debate is part of a larger one about the future direction of the American economy and society. As such, it is often diffuse and, at times, contradictory. It seems clear, however, that the debate will not only continue well into the future, but will also assume an even greater importance.

Pension funds represent a massive source of savings in our society. They provide a large and growing part of the capital resource base of the American economy. The question that we should pose is this: How should the deferred wages of working people be invested? What kinds of factors should be taken into account when those investment decisions are being made? Should the impact, both positive and negative, of those investments on the current and future economic security of pension fund owners play any part in that process? To that last question, investment professionals would resoundingly answer, "No!" They would say that their decisions are merely technical. They're responding to a market that is efficient. Investment professionals would say they are neutral and are simply being guided by the "invisible hand" that is the market system itself. But in an era when priorities are being set, when societal resources are being hotly contested, many others would answer that question with: "Absolutely, yes!" The people who are the owners of those funds should know that their assets are being used in ways that benefit them in the present *and* in the future, and that they are certainly not being undermined by their own money.

But the next, and much more difficult, question is this: How? It is easy to decide that something is wrong. It is much more difficult to determine what to do about it. We are only beginning to understand how to control these resources in ways that best represent the interests of the pension participants and their communities.

Ultimately, an alternative vision for the investment and control of workers' pension funds will also require the development of a strategy for organizing and allocating societal resources. This essential strategy for "economic organizing" must become a central part of the movement to gain greater and more democratic control over our resources.

Appendix

Table 4-1. Institutional Investors as Major Investors in "Fortune 500" Companies

*Institutional Investor**	*No. of Cos. Held*	*Stockholder Rank*					*Employee Benefit Funds as % of Total Assets Managed†*
		1	*2*	*3*	*4*	*5*	
J. P. Morgan & Co.	104	13	26	26	24	15	62%
Batterymarch Financial	70	2	14	24	19	11	87%
Capital Group	61	5	22	17	10	7	67%
Prudential Insurance	50	9	12	10	10	9	60%
TIAA-CREF	50	6	8	13	13	10	100%
Manufacturers Hanover	48	7	9	11	10	11	55%
Marsh & McLennan	42	8	8	7	13	6	67%
Citicorp	39	5	6	4	11	13	32%
Bankers Trust	35	9	7	9	5	5	74%
Delaware Management	34	6	16	3	6	3	90%
T. Rowe Price & Co.	30	8	6	8	4	4	58%
Fayez Sarofim & Co.	30	5	6	9	3	7	82%
Lord Abbet & Co.	29	4	4	11	5	5	66%
Mellon National	25	5	6	6	4	4	67%
FMR	25	0	4	5	6	10	90%
Pioneer Group	24	5	3	6	3	7	45%

Chemical Bank	24	3	2	9	6	4	58%
National Detroit	21	6	5	6	1	2	67%
State Street Research & Management	21	6	5	6	1	2	79%
Thorndike, Doran, Paine & Lewis	19	1	4	4	5	5	37%
Bank of America	18	2	5	2	5	4	73%
Chase Manhattan	17	2	1	2	6	6	47%
California Public Employees and teachers Retirement system	16	0	4	3	4	5	100%
Wisconsin Investment Board	16	0	2	5	3	6	82%
First National Boston	16	0	5	4	4	3	65%

*"Institutional Investors" are the multi-billion dollar money managers which specialize in investing other peoples' money. They manage the assets of wealthy individuals, corporations, and pension and other benefit funds. Below is a listing of institutional investors who rank among the top five stockholders in the Fortune 500 corporations.

†Bank trust percentages based on year-end 1979 figures. Other percentages based on year-end 1980 figures.

*Percentage of equity holdings.

SOURCE: *Fortune 500 Stock Ownership Directory*, Corporate Data Exchange, Inc.; Federal Financial Institutions Examination Council; *Investment Age*, March 30, 1981; *Money Market Directory*, 1981.

Table 4-2. The Growth of Pension Assets, 1955–1980

	Figures in Billions of Dollars, Year End					
	1955	*1960*	*1965*	*1970*	*1975*	*1980*
Private non-insured pension funds	18.3	38.1	73.6	110.4	146.8	286.1
Insured pension funds	11.3	18.9	27.3	41.2	72.2	164.6
State and local government pension funds	10.8	19.7	34.1	60.3	104.8	202.7
TOTAL PENSION ASSETS	40.4	76.7	135.0	211.9	323.8	653.4

SOURCE: Federal Reserve System, flow of funds, December 1976, February 1981.

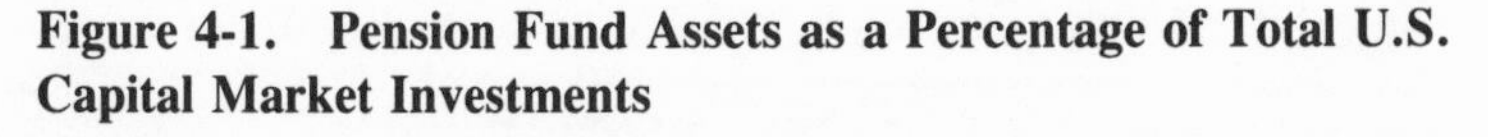

Figure 4-1. Pension Fund Assets as a Percentage of Total U.S. Capital Market Investments

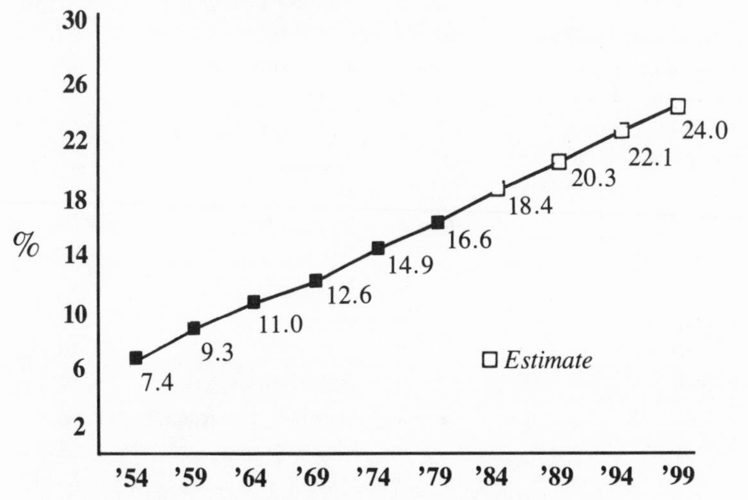

SOURCE: Federal Reserve System, Flow of Funds, November 1980.

NOTES: The chart graphically illustrates how pension funds have grown in importance to the American economy. In 1954, according to the Federal Reserve System, pension fund assets represented 7.4% of *total* U.S. capital market investments—all stocks, bonds, mortgages and government securities in the economy. By 1979, that proportion had more than doubled to 16.6%. Based on this trend, pension assets are expected to represent about one fourth of all capital market investments by the turn of the century.

U.S. Capital Market Investments include corporate equities, corporate and foreign bonds, U.S. Government bonds, federal agency securities, state and local government securities, and mortgages.

Table 4-3. Companies More Than 50% Owned by Institutions*

Northwest Air	75.3%	Florida Power	57.5%
Digital Equipment	74.7%	Continental Illinois	57.5%
Kellogg	74.1%	J. P. Morgan	57.4%
American Broadcasting	69.4%	Texas Utilities	57.3%
Hughes Tools	67.2%	Columbia Broadcasting	57.3%
Burroughs	66.3%	American Hospital	57.1%
Connecticut General	66.3%	St. Paul Companies	57.1%
Eli Lilly	66.2%	Bristol-Myers	56.8%
R. H. Macy	66.0%	Emerson Electric	56.8%
Cox Broadcasting	63.6%	Johnson and Johnson	56.3%
McDonalds	63.5%	INA	56.1%
Halliburton	63.0%	Caterpillar	56.0%
Motorola	63.0%	Baker International	55.8%
Travelers	62.6%	Stauffer Chemical	55.4%
K Mart	62.4%	Citicorp	54.9%
Intel	62.1%	NCR	54.9%
Revlon	61.9%	Abbott Labs	54.7%
Campbell Soup	61.9%	Pfizer	54.6%
Smith Kline	61.1%	Schering-Plough	54.5%
Texas Instruments	61.0%	Northwest Bancorp	54.5%
Raytheon	61.0%	Aetna	54.5%
Union Camp	60.8%	International Paper	54.4%
Deere	60.6%	Kimberly Clark	54.3%
Warner Communications	60.4%	Chesebrough Ponds	53.8%

Table 4-3. (*Continued*)

H. J. Heinz	60.4%	Clark Equipment	53.5%
Denny's	60.2%	Champion International	53.3%
Alcoa	60.1%	Lincoln National	53.0%
Associated Dry Goods	59.9%	Sperry	52.8%
Xerox	59.6%	Ingersoll Rand	52.4%
Monsanto	59.7%	American Home Products	52.3%
American Express	59.3%	J Ray McDermott	52.2%
Merck	59.2%	Allied Stores	52.2%
Dresser	59.0%	Du Pont	52.0%
Philip Morris	58.8%	Federated Dep't Stores	51.9%
Baxter Travenol	58.8%	Walt Disney	51.8%
General Mills	58.6%	Public Service Indiana	51.7%
Kaiser	58.2%	United Air Lines	50.9%
Atlantic Richfield	58.1%	Sterling Drugs	50.8%
Avon	58.0%	Searle	50.7%
Chubb	58.0%	J. C. Penny	50.6%

*Institutional ownership of selected corporations as a percentage of total outstanding stock, December 31, 1979.

SOURCE: *Institutional Holdings, December 31, 1979*; Stock Research Department, Salomon Brothers, April 29, 1980.

NOTE: Institutions include banks, mutual funds, investment advisors, private non-insured pension funds, state and local retirement systems, corporations, foreign institutions, endowments and foundations.

Table 4-4. Largest Pension Funds or Sponsors in the U.S.* as of September 30, 1980

Rank	*Name of Sponsor or Fund*	*Assets (millions of dollars)*
1	American Telephone & Telegraph Bell System Companies	31,100
2	California Public Employees	18,055
3	General Motors Corp.	14,123
4	Teachers Insurance Annuity Association/College Retirement Equity Fund (TIAA/CREF)	14,000
5	New York City Employees and Teachers	13,474
6	New York State Common Retirement Fund	13,308
7	New York State Teachers	7,914
8	New Jersey Division of Investment	7,570
9	General Electric Corp.	7,184
10	Ford Motor Co.	6,442
11	U.S. Steel & Carnegie Retirement Trust	5,540
12	Texas State Teachers	5,300
13	International Business Machines Corp.	5,200
14	Wisconsin Investment Board	5,015
15	E. I. DuPont DeNemours Corp.	4,930
16	Ohio Teachers Retirement System	4,742
17	Ohio Public Employees Retirement System	4,561
18	Michigan State Employees	4,559

Table 4-4. (*Continued*)

Rank	*Name of Sponsor or Fund*	*Assets (millions of dollars)*
19	North Carolina Public Employees	4,470
20	Exxon Corp.	4,339
21	Minnesota Investment Board	4,122
22	Florida Retirement System	4,000
23	Pennsylvania School Employees	3,968
24	Washington State Public Employees	3,126
25	Western Conference of Teamsters	3,110
26	General Telephone & Electronics	3,000
27	Sears Roebuck & Co.	3,000
28	Los Angeles County Employees	2,875
29	Shell Oil Co.	2,814
30	Standard Oil of Indiana	2,759
31	United Technologies Corp.	2,745
32	Rockwell International Corp.	2,730
33	Boeing Corp.	2,700
34	Eastman Kodak Corp.	2,653
35	Mobil Oil Corp.	2,623

SOURCE: *Investment Age*, January 19, 1981.

*Many funds and fund sponsors listed include a number of separate retirement plans which have been grouped together.

Bibliography

AFL-CIO, Building and Construction Trades Department. *Maximum Benefit: Making Investments Do the Job*. Resource Binder. Washington, D.C., 1980.

AFL-CIO, Executive Council, Committee on the Investment of Union Pension Funds. *Investment of Union Pension Funds*. Washington, D.C., 1980.

AFL-CIO, Industrial Union Department, Committee on Benefit Fund Investment Policies. *Pensions: A Study of Benefit Fund Investment Policies*. Washington, D.C., 1980.

———. "The Publication on Labor, Pension and Benefit Funds, and Investments." *Labor & Investments Monthly*, Washington, D.C., 1981.

Barber, Randy. *Pension Funds and Renewable Energy*. Washington, D.C.: Jobs in Energy Project, National Committee for Full Employment, 1981.

California Governor's Public Investment Task Force. *Interim Report*. Sacramento: Governor's Office, 1981.

———. *Final Report*. Sacramento: Governor's Office, 1981.

Chase, Thomas H. "Pensions and Politics." *Labor Issues of the '80s*. Basking Ridge, N.J.: AT&T Corporate Planning Emerging Issues Group, 1980.

Coltman, Edward, and Metzenbaum, Shelly. *Investing in Ourselves: Strategies for Massachusetts*. Washington, D.C.: Conference on Alternative State and Local Policies, 1980.

Conference on Alternative State and Local Policies. *Public Employee Pension Funds: New Strategies for Investment*. Washington, D.C., 1979.

Corporate Data Exchange, Inc. *Pension Investments: A Social Audit*. New York, 1979.

———. *Fortune 500 Stock Ownership Directory*. 1981.

Council on Economic Priorities. *A Study of Investment Practices and Opportunities: State of California Retirement Systems*. New York, 1980.

Drucker, Peter. *The Unseen Revolution: How Pension Fund Socialism Came to America*. New York: Harper & Row, 1976.

Employee Benefit Research Institute. *Should Pension Assets Be Managed for Social/Political Purposes?*. Washington, D.C., 1980.

Finkelstein, Mark A. *The Michigan Pension Project: A Working Paper on the Investment of Michigan's Pension Dollars*. Ann Arbor: University of Michigan, 1980.

———. "Public Employee Pension Funds and Economic Development." Ann Arbor: Honors Thesis, University of Michigan, 1981.

Greenough, William, and King, Francis. *Pension Funds and Public Policy*. New York: Columbia University Press, 1976.

Hansen, Derek. *Banking and Small Business: New Directions in Enterprise Development*. Washington, D.C.: Council of State Planning Agencies, 1981.

Harbrecht, Paul P. *Pension Funds and Economic Power*. New York: Twentieth Century Fund, 1959.

Harrington, John. *Packaging Housing Loans: Strategies for California*. Washington, D.C.: Conference on Alternative State and Local Policies, 1979.

James, Henry. "How Pension Fund Socialism Didn't Come to America." *Working Papers for a New Society* 4, no. 4 (1977).

Kieschnick, Michael. "Investing in the Public Interest: The Case of Public Pension Funds." *Working Papers*. Washington, D.C.: Council of State Planning Agencies, 1979.

———. *Venture Capital and Urban Development*. Washington, D.C.: Council of State Planning Agencies, 1979.

Kirschner, Edward, et al. "Public Pension Funds as a Source of Job Creation." California Employment Development Department, 1975.

Leibig, Michael. *Social Investments and The Law: The Case for Alternative Investments*. Washington, D.C.: Conference on Alternative State and Local Policies, 1980.

Litvak, Lawrence. *Pension Funds and Economic Renewal*. Washington, D.C.: Council of State Planning Agencies, 1981.

Litvak, Lawrence, and Daniels, Belden. *Innovations in Development Finance*. Washington, D.C.: Council of State Planning Agencies, 1979.

Mares, Judith. "The Use of Pension Fund Capital." *Working Papers*. Washington, D.C.: President's Commission on Pension Policy, 1979.

Messinger, Ruth, and the Municipal Finance Research Institute. *Revitalizing New York City's Economy: The Role of Public Pension Funds*. Conference on Alternative State and Local Policies, 1980.

New York State Assembly, Committee on Governmental Employees. *Public Employees Pension Funds: A Path to Increased Worker Benefits and Economic Revitalization*. Albany, 1980.

Parker, Richard, and Taylor, Tamsin. *Strategic Investment: An Alternative for Public Funds*. Washington, D.C.: Conference on Alternative State and Local Policies, 1980.

Pensions & Investment Age. Crain Communications, bi-weekly, New York and Chicago.

Petersen, John and Schotland, Roy. "Socially Useful Investments by State and Local Pension Funds." Washington, D.C.: Draft report, Municipal Finance Officers Association, 1979.

Petersen, John, and Spain, Catherine. *Working Papers on Public Pension Fund Investments and Disclosure Practices*. Washington, D.C.: Municipal Finance Officers Association, 1981.

Rifkin, Jeremy and Barber, Randy. *The North Will Rise Again: Pensions, Politics and Power in the 1980's*. Boston: Beacon Press, 1978.

Rosen, Kenneth. *The Role of Pension Funds in Housing Finance, Working Paper* No. 35. Cambridge, Mass.: Joint Center for Urban Studies of MIT and Harvard, June 1975.

Smart, Donald A. *Investment Targeting: A Wisconsin Case Study*. Madison: Wisconsin Center for Public Policy, 1979.

Tell, Lawrence. "Public Pension Investments Drain Millions from City Economy" and "From VWs to Kansas, 'Targeted' Investing

Yields Local Gain, No Peril to Pensions." *The Chicago Reporter*, July 1980.

Tilove, Robert. *Public Employee Pension Funds: A Twentieth Century Fund Report*. New York: Columbia University Press, 1976.

Triplett, Thomas J., The Minnesota Project. *Investing in Minnesota: A Proposal to Use State Moneys for Maximum Benefit.* Washington, D.C.: Conference on Alternative State and Local Policies, 1980.

Tropper, Peter, and Kaufman, Anne. *Pension Power for Economic Development.* Washington, D.C.: Northeast-Midwest Institute, 1980.

U.S. Congress, Senate, Subcommittee on Antitrust and Monopoly. *Beneficiary Participation in Private Pension Plans*. Staff Report, 96th Congress, 1st Session, 1979.

———. *Hearings: Pension Fund Investment Policies*, 2 vols. 96th Congress, 1st Session, 1979.

Vaughn, Roger. *State Taxation and Economic Development.* Washington, D.C.: Council of State Planning Agencies.

———. *Inflation and Unemployment: Surviving the 1980s*. Washington, D.C.: Council of State Planning Agencies, 1979.

Chapter 5

The Religious Community and Economic Justice

Rev. Charles W. Rawlings

The economic changes now unfolding in the United States and most other countries may be compared in magnitude and destructive character to the upheavals of the Industrial Revolution of the nineteenth century. Emma Rothschild has vividly captured the nature of the structural economic shift regarding three industries: eating and drinking places (including fast food), health and business which have

> accounted for more than 40 percent of the new private jobs created between 1973 and the summer of 1980. In that period their employment increased almost three times as fast as total private employment, and sixteen times as fast as employment in the goods-producing or industrial sector of the economy. . . . The *increase* in employment in eating and drinking places since 1973 is greater than total employment in the automobile and steel industries combined. Total employment in the three industries is greater than total employment in an entire range of basic productive industries: construction, all machinery, all electronic equipment, motor vehicles, aircraft, ship building, all chemical products, and all scientific and other instruments.[1]

The immediate human consequences can be measured in the tidal waves of job seekers. In Memphis, Tennessee,

40,000 people applied for 800 jobs at a new Japanese electronics plant.[2] 12,000 turned out in Toledo for 80 jobs.[3] In stricken Youngstown 1300 persons applied for 50 minimum wage jobs at Bob Evans Farm Sausage.[4]

Other essays in this volume document the magnitude of change more completely. However, the steel industry provides us with lessons we need to learn both about the scale of change and the moral darkness into which we have been invited. In the past three years 45,000 basic steel jobs have been lost through a process of disinvestment and the shut down of technologically uncompetitive mills. In the past ten years the job loss is more than 100,000 in basic steel.[5] Such figures must be multiplied, for each loss in basic steel, by an uncertain factor of two to four additional jobs lost in allied fields. The shock waves of a depreciating economy range from personal tragedy as families, homes and expectations are dashed and broken, to closed schools and hospitals where striking workers are discovering their derivative economy will no longer maintain their standard of living.

A deeper, and perhaps unintended, illumination of this process was provided by a recent issue of *U.S. News and World Report*, whose cover story about the comeback of the Marines was located alongside a splashy report on Pittsburgh's Renaissance II. The end of the city's reputation as a steel-making center was heralded with headlines about the transition from blast furnace to office towers. While the Marines were pictured in snow, tropical jungles, and sandy desert as the nation's new rapid deployment force to secure "America's global security interests," Pittsburgh was shown as a center for the management of global financial and corporate enterprises.[6]

U.S. News was optimistic about the future of the workforce in this transition, but Pittsburgh Mayor Calaguiri told the *Los*

Angeles Times he was aware that available jobs would decrease and offered the consolation that he would "rather have less people with high income, than more people with relatively low spending power."[7] In fact, it does not seem an unwarranted inference to conclude that Pittsburgh's freshly unemployed steelworkers will be expected to offer their sons—and maybe their daughters—to military service whose objective may be the relocation of American jobs in Third World countries. New steel mills in South Korea, Taiwan and the Phillipines have, in fact, already been built.

If these sons survive their encounters with regional resistance to American hegemony they may well return to discover not only their parents' declining economic prospects, but also a fierce competition for the smaller pool of well paid jobs in their town's new office towers. The long lines in Memphis, Toledo, and Youngstown already testify to this possibility.

Apparently the prophetic chapter in Barnet and Mueller's *Global Reach: The Power of the Multinational Corporations* entitled "The Latinamericanization of the United States" is being fulfilled. The movement of capital from industrial cities not only closes mills and factories but also creates a new and exploitable supply of cheap labor in the abandoned community. That is what the long lines for minimum wage jobs at the Bob Evans restaurant mean. Therefore, it will be crucial in the coming months to ask whether the various proposals one hears for reindustrialization have in mind breaking the backs of unions and the high standard of wages and benefits they have achieved through years of social struggle. Do such proposals include a return to the harsh days of an earlier era when workers were hired by industrialists to break each other's backs? The religious community must inquire responsibly, but with moral zeal, whether reindustrialization means Latinamericanization. More than that, the churches and synagogues of the nation must play a unique role as advocates for a justice that can reach its highest achievements only if it recognizes the priceless worth of each individual whom God has created and endowed in His own image. That justice needs an urgent and fresh translation in today's critical situation.

There are four priorities to be pursued by religious people. The first is to confront the silence about what is happening. Ending the silence about a process that is demonstrably destructive to people, their families and communities means taking the second step of legitimating their anger. In the name of social control, order, and responsibility, we have all been drawn into a dangerous and destructive process that involves choking off the very act of human integrity that by its outrage empowers people to take action on their own behalf. Without an end to the silence, the cover-up, the threshold step for democratic empowerment and action cannot be taken through a legitimate anger about what has happened.

A third task requires the churches and synagogues in America to learn to live alone. Like the divorcee or young woman or man reaching maturity and receiving the advice of good counselors, or good feminists who first must know their own roots, their own selves before entering into alliances with others, so the churches must learn to stand alone in the ideological struggles of left and right. In particular that separateness means the churches must sever their comfortable and historic relationship with capitalism. Our biblical roots in the justice teachings of the prophets and apostles have their own authentic message apart from any socialism or capitalism.

Finally, the religious community in America must help people rediscover a sense of transcendence, a sense of the wholeness, the creative possibilities in life. A society that purchases folk music idioms to sell savings accounts, makes masculinity dependent on car ownership, and names the place of women with lotions and perfumes has destroyed the possibilities for wholeness in life and society. Without the proclamation of that vision the other steps can come to no purpose.

Returning to the first theme, involving the question of silence and the need to end it, two stories illustrate a psychological and social aspect of the prevalence of silence. In the struggle of the churches in Youngstown, Ohio, in the face of the steel shutdowns, the need to maintain credulity was pervasive. Here was a company town that had operated within the shadow of Big Steel all its years. As strained as the relationships between labor and management were over those years, the town felt the company to be a source of nourishment. It was hard to stop believing in the nourishing aspect of the company. The underlying spiritual problem created a high degree of passivity captured in a vivid statement by a union local president who sought to explain why he and his men did not begin to take defensive action in their own steel mill after the giant Campbell works down the river had closed abruptly. In spite of the fact that warning signs and company signals were like handwriting on the wall, the men had stayed to themselves and remained relatively complacent. Later, after they were shut down, the union leader said "the plant is so huge, it's so big and the machinery is so enormous, when you work in it you can't believe it will ever close."

The statement is a paradigm for our national twentieth-century credulity in things big, technological and powerful. This union leader's insight tells all of us about our habits of spiritual dependency and our idolatrous relationships. The failure of the symbols of power is too ominous for a people who have grown psychologically and spiritually dependent upon them. The consequence is our silence even as disaster unfolds around us.

A second story gives additional shape to the silence that most readers will find familiar. A steel manager in Youngstown, much respected (although sometimes grudgingly) by the workers, helped the Ecumenical Coalition in the last phases of its planning for a reopened and modernized steel mill under community/worker ownership. He participated in discussions with the international union that reached agreement about the basic costs that could be calculated for labor in the reorganized mill. He gave useful counsel and advice about the plant's history and performance possibilities. In the process he reconfirmed the now familiar story of a plant that had always been profitable turned into a "cash cow" by a conglomerate, the Lykes Corporation, that wanted the cash flow for other purposes. The Youngstown Sheet and Tube Company had first turned its profits away to build a new mill on the Great Lakes at Indiana Harbor, then sold what was a home-owned business to shipping people in New Orleans.[8] The people had lost control of their own plant and their own capital in that process.

At the crucial moment in the effort to gain federal support for the plan to reopen the Campbell Works, the steel manager declined to give the very proposal on which he had worked his public endorsement. He was not *sure* it would work. But beyond that, he seemed caught in the strategy of his peers in the industry. The Lykes Corporation's closing of the giant mill in Youngstown was considered one of the most indefensible shutdowns by a company because of its long profitable record. Nevertheless, United States Steel immediately paid for large banners proclaiming "Foreign Steel It's a Job-Robbing Deal" which were hung by the city from nearly every public building in Youngstown.[9] The American Iron and

Steel Institute launched a campaign to tell the public that this and other mills would close because of government regulation of prices, EPA restrictions, and unfair foreign competition. There was virtually no mention of mismanagement that had caused the neglect of plant stock and equipment; no mention of the decades of uninspired technology in American steel while Japanese and German manufacturers modernized; no mention of the now well-documented pattern of disinvestment and diversification by Big Steel that caused the death of many plants.[10]

Enormous effort was expended by the steel industry to put a bright face on the first wave of shutdowns. The clergy were alternately branded naive or "red" because of their noises to the contrary. They received little support even from their parishioners. The high point in the campaign to keep silence about the destructive consequences of these policies by steel corporations came through that predictable conduit, *Fortune Magazine*, which went to bed at Thanksgiving 1979 with a story about the revival of Youngstown that could not be deleted in time to avoid its unfortunate coincidence with U.S. Steel's post-Thanksgiving announcement to close the last of Youngstown's integrated steel mills, pushing basic steel unemployment there to well over 12,000 in less than three years.[11]

The steel manager who helped the Ecumenical Coalition knew all these things. But he could not escape the "club" pressures to maintain credulity in the face of disaster. The industry line was that the shut downs were necessary; nothing else could happen; nothing else should. In the end the local manager chose to remain silent. He must be credited with holding a skeptical view of the reopening plan on honest grounds, but his silence seems still another parable of the problem and nature of silence about grave human disasters.

As we increasingly sense that something very terrible is happening to the nation's steel, rubber, auto, electronics and machine tool makers, we must become increasingly cogent about the underlying human question: what has happened to Youngstown's 12,000, Toledo's 8,000, Akron's 35,000, Detroit's 100,000? We must resist the temptation to be good Germans, maintaining our credulity and our religious isolation while the people perish.

An important footnote must be added to the problem of silence. This concerns the technical arguments that are used by economists, engineers and managers to silence critics or alternative proposals. Perhaps the most dishonorable aspect of the steel industry's attack on the Ecumenical Coalition proposal for community/worker ownership of a modernized and reopened mill was the use of technical arguments. The point is not that some issues are not complex; nor is it that good and moral intentions can ever compensate for lack of practicality. The problem is with the use of the Nixonian argument that only specialists have enough information to make a decision. In the case of Youngstown, it happens that the Ecumenical Coalition got good technical advice. The modernization proposals advanced both on behalf of the Campbell Works and, later, the Ohio Works conform precisely to the recommended technologies subsequently verified in studies by the Office of Technology Assessment, articles in *The Wall Street Journal* and *Fortune*.[12] The Coalition may have erred most in its estimate of the government's readiness to provide loan guarantees and other supports needed to replace the equity withdrawn from Youngstown by private capitalists. On either the technical questions, where the Coalition appears to have been correct, or the political question, where it appears to have been badly misled by the government and its own estimates, silencing constructive and responsible efforts is a form of political intimidation. The fact that the business community stooped to such tactics should not deter religious people from strengthening their readiness to accept risky and technical problems in the process of building a just economic and social order. Silence, whether in the name of complexity or illusion or powerful sectors of self-interest, is morally intolerable for a society committed by constitutional and religious values to seek and champion the commonweal. When the public good is injured on a mass scale, as in the case of plant closings, indignation and anger are appropriate to the creation of energy to restore justice.

It is exactly the absence of anger in many instances of plant closings that creates the passivity, the sense of hopeless acceptance, of the fate handed down by those in power. Perhaps the conflict-laden days of the 1960's and the harsh struggles to end the Vietnam War have led the churches and synagogues to emphasise their role as reconcilers and peacemakers. Yet, just as peace without justice cannot be called peace, so power without a legitimate anger will not provide victims of social injustices such as plant closings with the reality of their own experience.

There is no more urgent task before the churches and synagogues today than to legitimate anger about the destructive experiences people have. It is the only appropriate antidote to the sense of guilt taught in American culture as the partner of misfortune. The very absence of a sense of structural power and structural evil leaves individuals with only themselves to blame for bad news. The decision to close steel mills was very bad news for the workers; the same was true in auto plants, rubber factories and electronics plants. Yet in most instances scapegoats were the order of the day. Either the workers had become careless in their crafts; or, they had demanded too much money; or they simply were in old plants

Delaware Valley Coalition for Jobs

Delaware Valley Coalition for Jobs

Delaware Valley Coalition for Jobs

Unitarians, Quakers, and Jews in the "March to Save Our Jobs," Philadelphia, June 17, 1981.

located too far from water transportation; or it was the evil government, or the evil foreign imports. Since few workers can imagine themselves having impact on either their own or foreign governments, the psychological fate of unemployed, skilled workers has been to have no legitimate basis for their anger and no legitimate object toward which to direct it.

On human grounds alone the religious community must lend legitimacy to the anger of those injured. Just as important, practically, the religious community must help victims of injustice find the appropriate substance and understanding of their affliction.

This is not to argue for absence of an appropriate sense of proportion. Nor is it to call for destructive or riotous anger. But we must be aware that anger made legitimate only when it is turned on one's self is self-destructive. The problem for the victim of plant closings today is to avoid exactly that self-destruction. It is the self that is at stake regardless of the glosses put on events by all others. Accordingly, it should be the self that seeks the external causes of its injury. Certainly such taking stock may include some self-assessments that may be painful to face. But appropriate anger at injustice must include an angry assessment of those external forces or people who have done injury.

This predictably controversial role of the religious community has its origins in the function of anger in the biblical tradition. There it was commonplace for an indignant Moses, an outraged Jeremiah, and a furious Jesus. All the more striking in contrast is that city after city in America experiences the evisceration of its economy with such slight show of anger. The reasons go beyond the habits of self-recrimination to include the repression of anger. If anger only at oneself can be legitimate, it follows that those collective structures such as corporation or state that make destructive decisions from time to time have contrived the ideal defense; namely, that any collective anger is illegitimate. In this Orwellian sense, the collective entities in power can do no wrong. Their lies become truth; their wars become peace. People must take their pain, their injury and express themselves only to other individuals or visit it upon themselves.

In the late twentieth century the churches and synagogues have a mission to give people the legitimacy of their own anger at injury, both for the sake of their psychological and spiritual health, and for the sake of their possibilities for collective action for their communities and against those structures of state or corporate enterprise responsible for their suffering.

Precisely this reference to the problem of destructive policies in relation to either government or the private sector gives rise to our third task in the religious community today. That task is learning to live alone and apart from the ideological loyalties that have characterized much of our existence in the twentieth century. They are particularly true of the church's relationship to capitalism. The post-World War II era has given rise to the comfortable notion of a symmetry between the affluence of that period—and therefore its ideology, capitalism—and the church's mission. The past forty years represent the summit of the age of growth and its various phases: a westward movement, an industrial revolution, a technological era, an era of overseas colonial stakes. Each phase has lent itself to the notion that inequality, however omnipresent, was only temporary. Moreover, we must admit that that has been the case for significant segments of American population since the Great Depression. But it has also been an era of consistently marginalized minority and poor populations in the cities and Appalachian regions. Capitalism in Brazil and other Third World countries, while contributing a new affluent class of people, has increased the poverty of lower classes and led to even greater discrepancies in their incomes.[13]

Not only because of persistent inequality, but also because of the myths about growth and the promise of equality unrealized in the emerging age of scarce resources, the churches and synagogues must be clear about the justice questions which are their primary responsibility to raise in every concrete economic and social situation and in every ideological camp and political faction.

In order to live alone, the churches and synagogues must recapture their intellectual sophistication which has become lax in the age of happiness. The readiness of students emerging from theological seminaries is limited when we recognize that many seminaries in this country offer few or no courses on the social teachings of the church. Excellence in academic matters and relevance to the struggles of the contemporary human community are badly needed to prepare the congregations for their role in the desperate conditions of these last decades of the twentieth century.

Theologian Gibson Winter has said that the church is one of the few remaining free spaces in a society devoted to technology and power.[14] It is in that free space that we must help one another conceive a healthy and whole society from the fragments of destroyed people and communities that have become accepted public policy. In that same free space the religious community must join hands with others who share its vision of justice and caring. Foremost among those institutions is the labor movement. Joe Holland, the scholar working on issues of religion and labor at the Center of Concern, has reminded us that unions are one of the few places where people address each other as brother and sister.[15]

Yes, the churches must be discriminating about all institutions—including organized labor—that seek to become powerful influences in public affairs. But those institutions that embrace an egalitarian ideal stand in a special kinship to the biblical traditions of Jubilee, with its land redistribution and the Acts of the Apostles, where all things were held in common by the community. It is important for the religious community both to confess its easy alliance with affluent segments of American society and to rediscover the record of struggle—sometimes bloody struggle—in which working people in this country have engaged to mold a vision of a caring, loving, just society. Even as those working people are being eliminated from economically productive roles by vast industrial shutdowns, we must revive our understanding of their historic struggle as working people in a society that has honored wealth more than people and power more than justice.

For the religious community, learning to live alone means reclaiming its advocacy for creation and for the first fruits of creation: the women and men and children of our communities. The biblical admonitions about injustice, false gods and judgment derive from an adversarial relationship between prophets and the powerful. That heritage of faith calls us all to return to visions of a kingdom of justice and peace; one that shuns those venal, narrow, corrupting and murderous values that have been elevated to respectability in our present era.

Through its struggles in civil rights, to aid the farm workers, to end the Vietnam War, and to renew Youngstown, the churches have come to the possibility of a new maturity as they face the present crisis of economic injustice. As the religious community becomes intolerant of silence about such injustice, as it empowers the victims indignantly to seek their own justice, and as it remains faithful to its own identity with the people of God we shall have set the foundations for new social institutions and for new alliances with all those who address each other as sister and brother. We shall fulfill the words of the prophet Isaiah:

> Is not this the fast that I choose:
> to loose the bonds of wickedness,
> to undo the thongs of the yoke,
> to let the oppressed go free,
> and to break every yoke. . . . If you pour
> yourself out for the hungry and satisfy
> the desire of the afflicted. . . . Your
> ancient ruins shall be rebuilt; you shall
> raise up the foundations of many generations;
> you shall be called the repairer of the breach,
> the restorer of streets to dwell in.[16]

Notes

1. Emma Rothschild, *New York Review of Books*, Feb. 5, 1981.
2. *Wall Street Journal*, Feb. 26, 1981.
3. *Cleveland Plain Dealer*, March 4, 1981.
4. *Warren* (Ohio) *Tribune*, March 18, 1981.
5. *Wall Street Journal*, April 2, 1981.
6. "Pittsburgh's Renaissance," *U.S. News and World Report*, Jan. 12, 1981
7. Quoted by Msgr. Charles O. Rice in the *Pittsburgh Catholic*, Nov. 21, 1980.
8. *Conglomerate Mergers—Their Effects On Small Business and Local Communities*, Hearings before the Subcommittee On Antitrust and Restraint of Trade Activities Affecting Small Business of the Committee On Small Business, House of Representatives, Jan. 31–Feb. 28, 1981, pp. 57–90.
9. Reported by Mayor Philip Richeley, Youngstown, Ohio.

10. "Shut Down and Destroy: The United States Steel Corporation Versus the Public Interest," published by the Tri-State Conference on the Impact of Steel (400 Hoodridge Drive, Pittsburgh, Pa., 15234), April 1980, pp. 3–4.

11. "Youngstown Bounces Back," *Fortune*, Dec. 17, 1979.

12. *Technology and Steel Industry Competitiveness*, Office of Technology Assessment, Congress of the United States, June 1980, Library of Congress Catalog No. 80-600111; *Wall Street Journal*, April 2, 1981; "Big Steel Recasts Itself," *Fortune*, April 6, 1981.

13. Penney Lernoux, *Cry of the People: The Role of the U. S. Government in the Rise of Fascism, Torture and Murder and the Persecution of the Roman Catholic Church in Latin America* (New York: Doubleday, 1980).

14. Gibson Winter, Address at the North/South Institute, College of Preachers, Washington, D.C., March 31, 1978.

15. Joseph Holland, Address at the National Conference On Religion and Labor, May 13, 1980, at Thomas More College, Covington, Ky.

16. Isa. 58:6–12 (RSV).

Chapter 6

How to Civilize Capital

Robert Lekachman

I start with praise of Calvin Coolidge. When he cogently remarked that the business of America was business, he was entirely correct. With only rare and brief exceptions, our dominant interest group has been business, increasingly in the shape of the large corporation. As Yale University's born-again political scientist Charles Lindblom[1] pointed out in his magisterial *Politics and Markets*, the corporate sector is "more equal" than its rivals because it has two shots at shaping public policies and directing public opinion in its own interests: the conventional capacity to reward and frighten actual and potential officeholders that it shares with other organized lobbies, and, something much more potent, its role as keeper of the cornucopia. A conservative society, crippled by the weakness of a credible political left, perceives the private sector as the source of jobs, incomes, individual status, and advancement.

This concentration of ideological, political, and economic power goes far to explain the decision of a sufficient number of the voters in November 1980 to reward their tormentors with the Presidency, a majority of the Senate, and a working plurality in the House of Representatives. For the fact of the matter is plain—that the efficient causes of inflation, high unemployment, flagging productivity, and the decay of communities in the Northeast and the industrial Midwest are firmly located in the malfeasance, misfeasance, and nonfeasance of corporate America.

Let us remind ourselves of a few of the ways in which American business has propelled us into our current crisis.

- Failure to keep up with the state of the art in autos, steel, and consumer electronics. Even now the American position in computers and microchips is under assault by Japanese rivals. The highly overpaid leaders of Detroit's auto industry stubbornly resisted the market implications of OPEC's 1973 coup and the sensible shift of their customers' preferences toward smaller, energy-efficient vehicles. As for steel, few American facilities, aside from a few fabricating specialty products, measure up to Japanese criteria of modernity.
- One would scarcely guess that statistical quality control is an American intellectual achievement, arrogantly neglected by American managers. Events have demonstrated that the customers are less stupid and less suggestible than the virtuosi of marketing and advertising black magic think they are. Presented with a choice between Japanese and American TV's, cameras, watches, and so on, they buy Japanese less for price and more because the foreigners furnish more reliable products of generally higher quality.
- OPEC is a convenient scapegoat for soaring energy prices, but the roots of our dependence on imports are imbedded in the policies of the Seven Sisters, Exxon and its gigantic colleagues, who promoted wasteful use, destroyed mass transit in cities like Los Angeles, and in the 1950's, pushed through Congress an interstate highway system which to this day subsidizes automobiles and trucks while penalizing railroads and mass transit.[2]
- Incessant complaints from business quarters about the costs and burdens of environmental controls, regulation of occupational safety and health, increased product reliability, and so on ignore the blatant fact that these regulations represent a belated and inadequate response to the callous and heedless damage inflicted

upon men and women and their natural environment by businessmen fixed catatonically on their bottom lines. Until very recently they lowered their own costs of operation by imposing heavy medical and community costs upon others. I refer here to what economists delicately term externalities.[3]

- Much of the reason why defense is so expensive relates to the gross inefficiency of the Pentagon and its satellite contractors, signaled for many Americans by the tragic failure of the April 1980 attempt to rescue American hostages in Iran.[4]

Liberals and radicals are not alone in their criticism of recent corporate performance. As thoughtful business commentators in sympathetic journals like the *Harvard Business Review* and *Business Week*[5] have come to recognize, corporate leaders have been so obsessed with immediate profit, so little disposed to look ahead, so selfishly motivated to advance their own careers by jumping from one corporate employer to another, and so neglectful of foreign markets and the inroads in domestic markets of Japanese and European rivals, that they have neglected quality, skimped on research and development, and arrogantly ignored the preferences of their customers.

As a national leader, President Carter lamentably failed to identify the sources of economic malaise. Indeed, in the desperate last year of his term, he essentially adopted Republican responses to inflation: he initiated the short but frighteningly sharp recession which occurred in the second quarter of 1980. Carter's lack of coherent policy and effective articulation consigned him and his party to a thoroughly merited defeat. The tragedy of the 1980 election was not that a weak, confused, and conservative chief executive lost; it was that the only alternative, absent a credible candidate on the political left, was Ronald Reagan, the most reactionary presidential candidate offered by a major party in this country since Herbert Hoover.[6]

I

Needless to say, my analysis is far different from that of Reagan ideologues. They appear genuinely convinced that contemporary deficiencies in the performance of the American economy derive not from managerial shortcomings but from overspending on social programs, coddling of the lazy, bureaucratic meddling, excessive regulation, and supposedly oppressive taxation.[7] During the interregnum between the November 1980 election and the inauguration of a new president in January 1981, then Congressman David Stockman[8] and his House colleague Jack Kemp composed and presented to the president-elect a document which at least in the initial phase of the Reagan administration has served as a blueprint. Entitled "Avoiding a GOP Economic Dunkirk," the polemic possesses the merit, if none other, of candor. Under five headings it details an agenda startling in its anachronistic echoes of 19th century individualism and the social Darwinism which legitimated it.

It is worth detailing the specific recommendations because, regardless of their ultimate disposition by Congress, they are the heart's desire of this administration.

"Supply-side Tax Components"

In pristine form, all that supply-side economics amounts to is the proposition that lower taxes will stimulate saving and investment, encourage professional and skilled men and women to work harder, and persuade entrepreneurs to gam-

Barbara Baker: Campaign for Human Development, U.S. Catholic Conference—used with permission

Unemployed workers lend their support at the Eaton demonstration, May 18, 1981.

ble on new processes and products in the hope of larger gains.[9]

It follows that the best way to improve life for ordinary working slobs and the genuinely poor is to encourage the rich and the enterprising to do their thing. Hence Stockman-Kemp (hereafter S-K)[10] recommend immediate Congressional passage of the first of three ten percent reductions in personal income tax levies. They urge in addition reduction of the current seventy percent top rate on dividends, interest, and other property income (paid by few and readily avoided with the help of astute accountants and tax lawyers), and "a substantial reform" of corporate depreciation to the end of enlarging profits. The undisguised, indeed eagerly proclaimed, object of these exercises is more profit, fatter dividends, and a substantial shift from wages and salaries to property income.

"Fiscal Stabilization"

Here is a modest budget-cutting list to make the hair curl. Read closely to find your favorite programs:

- Cut by ten to twenty percent funds for roads, mass transit, sewer facility treatment, public works, national parks, and airports.
- Reduce by various percentages such entitlements as food stamps, cash assistance, medicaid, disability, heating aid, housing rental supplements, school lunches, and unemployment compensation.
- Curtail by a third, say $8 billion, "low priority" programs such as NASA, CETA, Urban Development Action Grants, the Community Development Program, the Economic Development Administration, urban parks, economic impact aid, Action, the endowments for the Arts and Humanities, and the Consumer Cooperative Bank.

– Cast a cold eye upon federal credits, loans, and loan guarantees. Here S-K temper their ideological zeal with a touch of vagueness. Possibly they reminded themselves that farmers, homeowners, small businessmen, developers, and credit-seeking parents of college students often vote Republican. It is ever so much easier for Reaganauts to be concrete and specific about food stamps and other aids to low income Americans who stayed home November 4, 1980, or unpatriotically voted for Democrats.

"Regulatory Ventilation"

S-K gravely warn against an "explosion" in the burdens of regulation to be imminently inflicted upon American business unless standards and regulations not quite in force are delayed, revised, or, best of all, junked. Most of the targets concern safety and fuel efficiency in new automobiles, industrial boilers, waste water, efficiency of appliances, Occupational Safety and Health Administration noise standards, similar OSHA limits on exposure to asbestos, cadmium, and chromium, and grain elevator dust control. S-K look with equal disfavor upon affirmative action hiring as still another erosion of individual liberty.

"Contingency Energy Package"

For solution to our nagging energy problem, S-K prescribe the usual magic. Before a new oil shock hits, the President should dismantle all controls; and, should the pinch become severe, he should boldly initiate a "package of emergency steps to increase short-run domestic energy production and utilization." The list is enough to make the strong blanch: quick licensing of half a dozen completed nuclear plants, removal of stripper and marginal wells from the reach of taxes on windfall profits, and "emergency variances" in standards for industrial and utility coal boilers.

"A Monetary Accord"

Had enough? Pause for S-K's final recommendation. After a meeting at the summit between President Reagan and Chairman Paul Volcker, the Federal Reserve, legally an independent agency, should interpret its new marching orders as "a new informal charter—namely, to eschew all considerations of extraneous economic variables like short-term interest rates, housing market conditions, business cycle fluctuations, etc., and concentrate instead on one exclusive task: bringing the growth of Federal Reserve credit and bank reserves to a prudent rate and stabilization of the international and domestic purchasing power of the dollar."

II

As social policy, this blueprint, faithfully followed in the early months of the new administration, is both calamitous and single-minded to the edge of obsession. The use of contemptuous language like "soup kitchens" to stigmatize the meager and incomplete shelters against illness, unemployment, structural shifts in the economy, and old age and disability which the rickety American welfare state grudgingly accords the unlucky, bespeaks the mood of the genuinely reactionary right. Echoes resonate of the social Darwinism much in vogue in the writings of Herbert Spencer a century ago. Let the strong flourish and the weak perish, for the improvement of the breed. Or as Thomas Robert Malthus compassionately noted at the end of the eighteenth century,

"At nature's mighty feast, there is no vacant cover for him." The "him" represents the large company of losers.

Only a trifle less blatantly, this document assaults trade unions and administers an unneeded boost for the open-shop, low-wage Sunbelt. It chimes melodiously with the McGill Commission Report[11] disclaimed by the president (Jimmy Carter) who appointed its members. To restrict unemployment benefits and curtail food stamps and medicaid is tantamount to increasing pressures upon laid-off men and women to move south and reconcile themselves to jobs in nonunion enterprises at much lower wages, or, if they remain where they are, flood depressed labor markets and further depress wage levels.

American national policy has been notoriously skewed against older cities and regions. Only the fatuous, cynical, or ideologically benighted still argue seriously that tax policy, subsidies to homeowners, interstate highways, federal grant formulas, and energy initiatives have been either fair or unrelated to regional competition. Limitation of social spending, energy deregulation, removal of social protections, shrinking of entitlements, and any projected monetary crunch certain to aggravate unemployment in the Northeast and industrial Midwest, do amount in combination to a program for the decline of older cities and regions, and a wager on a continuing and accelerating Sunbelt boom based upon available energy, cheap labor, and as close an approximation as can be arranged of the balance of power between workers and employers which existed during the Industrial Revolution. It is a vision of the near future which, it is almost redundant to add, rejects as a violation of natural liberty, the sanctity of contracts, and the sacred rights of property even timid efforts in the direction of social control of investment, restraints upon the flight of capital, rational restructuring of faltering industries, and equitable public controls over key prices and incomes.

III

I ask Lenin's question without adopting Lenin's answer: "What is to be done?" I do not underestimate the disarray of the political left, the general disaffection with public action to which that disarray is closely allied, or the power of emotion which propels the political right. On the last score, there is an especially alarming alliance between the Moral Majority and old-time economic fundamentalism.

All the same, I expect this configuration of political forces to be temporary, for the reason, above all others, that Reagan's policies are likely to be calamitous. A year or so from now, disaffected Democrats who voted for the California charmer will be sighing for the golden age of the Carter Administration. In particular, blue-collar workers who stayed home or voted for Republicans[12] will begin to comprehend better their situation and their interests.

Americans are pragmatists. They turned against the Carter Administration because it combined unacceptable quantities of inflation and unemployment and offered no credible prospect of doing any better if retained in power. They will reject Stockman-Kemp with as few qualms as their fruits turn bitter. The time is near when a credible program presented by the democratic left will constitute serious politics. Let me sketch the outlines of such a program, one which is hospitable both to local action and the need for coherent national responses to a bleak economic environment.

Jobs

Lying neglected and unused is the 1978 Humphrey-Hawkins Act, which, weakened and attenuated though it was by repeated compromise, nevertheless did improve upon the toothless 1946 Employment Act by setting specific but subsequently ignored unemployment targets and faintly sketching a planning mechanism. Meaningful committment to full employment is a genuinely radical social objective because it entails as corollaries several additional public interventions to which I turn next.

Restructuring of Major Industries

Even mainstream economists concede that on occasion markets fail. As a rough and ready proposition, private industries falter when suppliers mismanage the resources of capital and human labor they control and generate, as a consequence, prices so high that only a diminishing proportion of their potential customers can afford what they are selling. By this test, market failure is endemic in the American economy. The roster of ineptitude includes:

- *Health services*, which now consume an inordinate share of Gross National Product and channel an unduly large share of public expenditures into the purses of physicians, health insurers, hospital trustees, and other commercial vendors. The vast cost of medicare and medicaid has served as an argument for not moving toward the comprehensive health coverage routine in more advanced societies like Great Britain, Canada, Israel, and Sweden.
- *Housing*, an industry which on the record offers neither rental nor single family shelter for families of moderate income, let alone those worse off. A signal of its incompetence is the increasing number of communities which have recently adopted rent

controls. The spread of these controls should be a convincing response to those who impute New York City's scarcity of affordable housing to the continuation of local rent control.

- *Energy*. Sensible analysts such as Daniel Yergin and Robert Stobaugh[13] at Harvard's nonradical School of Business, Barry Commoner, and Amory and Hunter Lovins concur in their prescription of emphasis upon conservation and development of solar power. The financial interests of Exxon, Mobil, and their gigantic chums are precisely opposite. Each OPEC twist of the price screw additionally enriches them and their shareholders. They have craftily positioned themselves not in the widely dispersed, community-based solar experiments advocated by Commoner and Lovins but in the big solar of platforms and satellites which require so much capital that small and moderate sized operators are excluded.

These three industries, a large percentage of the American economy, are ripe candidates for public intervention, for creative experiments in local and cooperative ventures, and for democratic national planning.

Planning and Social Control of Investment

Here I reach directly the theme of this volume and the conference whose proceedings it embodies. The media, a huge majority of my colleagues, and, as a result, the general public routinely contrast planning and markets much to the advantage of the latter. As the usual argument runs, planners at best are clumsy and bureaucratic, and at worst arrogant and coercive. In inspiriting contrast, power in competitive markets is dispersed; and the most autocratic of chief executive officers are thwarted by equally authoritarian rivals and the capacity of the customers to spend their money elsewhere. The test of the market is harsh but equitable: the efficient survive and wax great, and the inefficient vanish from the scene. As the business advertisements occasionally emphasize, our system is a profit *and* loss system.

Daily experience is the best answer to the claims of efficiency made by apologists for American corporate enterprise. As for the argument against bureaucracy, it ought to be clear that bureaucracy is an unavoidable structural element of organizational scale, not the consequence of a distinction between public and private activity or ownership. Universities, public utilities, credit card issuers, major retailers, and large banks are as aggravatingly bureaucratic in their dealings with customers and employees as any local or federal agency. Those who genuinely seek less bureaucracy should be prepared to break up large corporations and decentralize public activity.

These, indeed, are objectives good men and women on the democratic left should be seriously pursuing. Concentrations of power, wherever located, are dangerous. The bad record of major industries dominated by small numbers of very large operators ought to suggest that even the economist's sacred cause of efficiency might be better served by small structures which widen the scope for human creativity.

I believe that the major challenge to the democratic left is a design of arrangements which at once guarantees broad social control over the investment process *and* dramatically decentralizes the exercise of such control. The fact that left liberals and unabashed radicals are only at the beginning of this vital task need not inhibit the definition of a set of interim principles governing the investment and disinvestment practices of major corporations.

At the least, we ought to require our corporate goliaths to supply,

- *Notice*: Save under the pressure of imminent (and independently verified) bankruptcy, enterprises which plan to close and reopen elsewhere must give at least six months notice to concerned constituencies of employees, suppliers, and local public officials.
- *Justification*: The enterprise should be required to demonstrate that any projected shift is not motivated by a preference for assorted tax and subsidy benefits or a yearning for a nonunion or antiunion environment. In other words, a corporation making adequate profits ought not be allowed to pursue inordinate profit against the public interest.
- *Compensation*: No corporation shall escape payment of severance allowances, relocation benefits to employees who choose to move with their employer, and lump sum compensation to communities who lose revenues from vanished property levies.

Corporations must be required to accept reasonable offers for facilities they propose to abandon, from communities, employees, or other noncorporate, interested parties.

One promising way to cope with the notorious drift of multinational operations to low wage, authoritarian, and non-union havens, is to impose penalty tariffs and quotas on merchandise which they seek to bring into the United States from such havens.

Incomes Policy

From conservative economists and business groups, the final argument against full employment is the threat of accelerated inflation. To this assertion, several responses are available:

1. Unemployment and underutilization of productive facilities themselves add to cost and thus to price inflation,

2. Tight money and punitive interest rates increase business costs and serve as an additional inflationary pressure. They also discriminate against small businessmen, family farmers, and state and local governments. They reward large corporate borrowers, which generates few jobs and fewer innovations,
3. Incomes policy does interfere with allocative efficiency but only when markets are genuinely competitive after the fashion of a diminishing number of American examples. Industries like steel and autos, which prefer high price-low volume marketing strategies to the low price-high volume alternative, actually misallocate resources by operating at inefficiently low percentages of capacity.

IV

No doubt these broad guideposts to democratic planning seem the most utopian of wishful thinking in this period of triumphant reaction. But let us be of good cheer. Our time will come; and although I embrace economic disaster no more enthusiastically than any other of its potential victims, I am confident of two things. We are heading for such a disaster. But out of it, if we do not flinch from the good fight, we shall pluck opportunity, enjoy renewed radical vigor, and fill that conspicuous gap in American politics—the gaping hole where a democratic party of the socialist left ought to be firmly rooted.

Notes

1. Lindblom in collaboration with Robert Dahl, who has also altered his views, was celebrated as a pluralist, an ingenious expositor of the view that competing interest groups tended roughly to cancel each other or at least veto the more outrageous proposals offered. Presumably the general interest emerged more or less unscathed from the rough and tumble of this struggle.

2. Consult on these matters the late John Blair's *The Control of Oil* (New York: Pantheon Books, 1976), and Robert Engler's *The Brotherhood of Oil: Energy Policy and the Public Interest* (Chicago: University of Chicago Press, 1977).

3. The Reagan administration's evident determination to loosen regulation will certainly imperil the health and degrade the amenities of numerous Americans. To the degree that it succeeds, it will reverse the wholesome extension of some social control over heedless pursuit of profit.

4. Inflation in the defense sector runs at approximately double the general rate. To the extent that the federal budget is skewed in the direction of military hardware, inflation is exacerbated.

5. See, for example, the June 30, 1980, issue devoted to "The Reindustrialization of America."

6. I may be doing Hoover an injustice.

7. George Gilder's *Wealth and Poverty* (New York: Basic Books, 1981) is the locus classicus of Reagan ideology. It is a fervent hymn to the power of avarice and a firm declaration that the poor need the spur of their poverty in order to gather their energy and become middle class like the rest of us. That the book has been a bestseller says something about the confusions of the public mind at the start of the Reagan administration.

8. Stockman became Director of the Office of Management and Budget and Kemp was elected by his colleagues to the number three position in House Republican leadership hierarchy.

9. See, in addition to Gilder, Jack Kemp's *American Renaissance: A Strategy for the Nineteen Eighties* (New York: Harper & Row, Inc., 1979), and Jude Wanniski's *The Way the World Works: How Economies Fail and Succeed* (New York: Basic Books, 1978).

10. See "GOP Plan to Avert 'Economic Dunkirk,' " *U.S. News and World Report* 89: 96–97.

11. See *Urban America in the Eighties: Perspectives and Prospects*, a report of the President's Commission for a National Agenda for the Eighties (Superintendent of Documents, Washington, D.C.). The flap over this document was a final example of the

Carter administration's notorious talent for shooting itself in the foot.

12. It is worth recalling that President Reagan's campaign appeal to blue-collar constituencies featured a promise to diminish unemployment and increase the number of new jobs.

13. See Roger Stobaugh and Daniel Yergin, *Energy Future: The Report of the Harvard Business School Energy Project* (New York: Random House, Inc., 1979).

Part Two

Philadelphia: The Deindustrialization of an American City

Chapter 7

Philadelphia: The Evolution of Economic Urban Planning, 1945–1980

Lenora E. Berson

"Philadelphia must create a climate in which business can flourish," Alan Brown, the city's first black Deputy Director of Commerce said, sitting in his sunny office on the sixteenth floor of the Municipal Services Building. "The business climate has not been good under previous administrations. It is our goal to prove that business can work in a cooperative nature with the city administration. It is our job to set the tone."

Looking out over the handsome cityscape, Brown was optimistic. "I do not think Philadelphia will be hurt by the recession. Our economy has shifted away from a manufacturing base to a service base. We have lost almost all of the jobs we can lose. Corporate planning is continuing, and we are in the process of plotting out where Philadelphia should be going."

The planning process includes the recognition that "it is very hard to convince corporate leadership living outside the city to move into the city or to keep plants now located in

Philadelphia anchored here. To a large extent we are at their mercy, so we have to target businesses."

Targeting, according to Brown, is a way of looking at businesses and deciding which are likely to survive, which can be expected to fail and which might be helped to expand. The result will be a strategy designed to stabilize weak industries such as the textile industry, help strong industries such as the printing industry to expand, and make it possible for new businesses to come into existence.

In pursuit of these goals the city will continue to offer marketing assistance as well as land and tax incentives to encourage large scale development. It will review its business taxing policies. The city will also take a long hard look at pollution controls of air and noise and their effect on business profit margins. In addition, new financing instrumentalities will be created.

Interestingly, most of the ideas expressed by Brown have been in circulation in Philadelphia since the fifties. The new addition to the mix is the special concern for small and neighborhood business.[1]

Brown, whose principal responsibility is the care and feeding of small business, explained, "Small business is the backbone of the Philadelphia economy. Over 85 percent of the city's firms have fewer than twenty employees. In addition these 27,000 firms represent more than 50 percent of the city's capital assets."

To aid small business, the city government hopes to set up a development bank. The bank will create a pool of venture capital. This money will be used to start new businesses and expand older ones.

"We expect the bank to be so profitable that its functions will be taken over by the private banking community. We must use the power of the city to get banks to lend money to small businesses. We must show the need and the fact that such loans are profitable. Banks, you know, do not create businesses. Rather they sustain them," Brown commented.

In addition, a city development corporation will offer a wide range of technical and managerial services to small businesses. Brown concluded by saying that he anticipates $1,300,000,000 of new development in Philadelphia. Most of this development he stated would come from "the tremendous growth of office space in Center City." As to jobs, he smiled. "I think employment will improve. The bleeding has stopped."

In addition to the items mentioned by Brown, the Green Administration has also committed itself to: revitalizing physically and economically moderate and low income neighborhoods in cooperation with existing neighborhood organizations; establishing programs to train the unemployed and underemployed; setting up an early warning system that would require plants planning to close or migrate to alert the city of their intentions; making Philadelphia an international city by revitalizing the port and creating an international trade zone similar to those existing in fifty-three other U.S. cities; luring foreign businesses to Philadelphia; expanding mass transit, improving its quality, keeping down the fares and helping establish small taxicab companies to augment existing services; maintaining federal governmental installations and obtaining federal contracts such as that of repairing the Saratoga; and working closely with the city's business leaders and their organizations.[2]

To meet the city's fiscal crisis, the Green administration plans to keep a lid on taxes and wages of municipal and allied governmental workers, eliminate patronage jobs and contracts, and present honestly balanced budgets.

The optimism expressed by these pledges is set against a somber historical background. In 1950 the population of the city peaked at 2,071,605. Every census since has shown an ever accelerating decline. The 1980 census reveals a citizenry of only 1,680,000. The population has not only grown smaller over the last three decades, but it has also grown older, poorer, and blacker. In addition, Philadelphia is now home to a significant number of low-income hispanics who immigrated here in pursuit of a better material life.

Since 1970 the city has lost more than 130,000 jobs, over 80,000 of which were in the manufacturing sector. Since 1950 employment has increased nationwide by 25 percent while the number of jobs in the city has declined by 8 percent.

In 1978 the city printed a monograph entitled *The City of Philadelphia: An Urban Strategy*. In a frank appraisal of the situation, it said: "In addition to absolute job losses, the structure of the economy has drastically changed. Since 1950 employment in construction has decreased by 35 percent, employment in manufacturing by 50 percent, and employment in wholesale and retail trade by 20 percent. Over 45 percent of all jobs in 1950 were in manufacturing; by 1978 that number had fallen to less than 20 percent. On the other hand, service employment had more than tripled. Unfortunately, these gains have only partially offset the employment losses occurring in other sectors of the economy.[3]

"These changes in the structure of Philadelphia's economy, the failure of the city's economy to keep pace with the nation, and the migration of jobs from the city to the suburbs have created significant problems. . . . First, the shift from blue collar manufacturing jobs to white collar jobs in services or finance, insurance and real estate places an enormous burden of readjustment on the city's labor force. In many individual cases, this readjustment is very difficult to make. Second, the dispersal of jobs throughout the region is particularly disadvantageous to minorities and low-skilled workers who are concentrated in the city and have limited means of transportation. The pool of available and accessible jobs for these individuals is continually shrinking. This has resulted in a severe unemployment problem for the city which hits particularly hard on the young and the minorities. The percentage of black male unemployment in Philadelphia is twice the city average, which itself is persistently above regional and national averages.[4] Further, the absolute numbers of unemployed are increasing despite a declining labor force. Finally, the lack of job opportunities for Philadelphia residents manifests itself in reduced personal incomes. Not only have the incomes for Philadelphia families been consistently below incomes in suburban counties, the growth of income . . . has been lower."

As this report, prepared by the Rizzo Administration (1971–80), indicates, awareness of the true nature of the city's unemployment problem predates the current Green Administration. In fact awareness of the consequences of shifts in the location and nature of employment goes back to the late forties and lies at the very heart of Philadelphia's famed reform movement (1951–62).

Since 1951 city officials have consciously been struggling with the consequences of Philadelphia's changing population. Over the years the city government has tried to maintain and expand its white working class population and its white professional and upper middle class. Only intermittent efforts have been made to upgrade the growing pool of low income black workers.

In many ways World War II serves as a watershed in Philadelphia history. Having met the challenges posed by industrialization and large-scale European immigration in the century before the war, the city found itself in the mid-1940's with a new set of problems. These difficulties can be categorized in four ways:

Public hearings on plant closings organized by the Delaware Valley Coalition for Jobs in Philadelphia City Hall, February 16, 1980. *Top left*: Congressman Robert Edgar, chairman, Northeast-Midwest Congressional Coalition, serving on the hearings panel. *Top right*: workers at ITE Gould plant in Philadelphia. *Bottom*: Fr. Joseph Kakalec, chairman, Philadelphia Council of Neighborhood Organizations, testifying.

1 The change in the economy, characterized by a shift from manufacturing to service jobs and by a movement of business and industry out of the city to the suburbs and beyond.
2 The physical decline of the city created by depression and war in the years 1929–45. During this period little new was built and little that was old was repaired.
3 The change in the population led by the establishment's move to the suburbs. Following them were large numbers of the children and grandchildren of the European immigrants. Replacing these groups in the city was a massive flow of poor unskilled rural southern blacks.
4 The inadequacy of the city's old fashioned political and governmental machinery to deal with these three problems.

To tackle these difficulties, a series of leadership groups with overlapping memberships emerged. The groups were:

- The Young Turks, comprised of idealistic professionals—lawyers, architects and city planners. For the most part these individuals were active in such organizations as the City Policy Committee, the Philadelphia Housing Association, and the Citizens Committee for City Planning. Each of these organizations immersed itself in detailed plans for renewing the city's fiscal, physical and political plant.
- The Business Establishment, led by such commanding old Philadelphia family figures as Walter Phillips, Sr. and Edward Hopkinson. Organizing into the Greater Philadelphia Movement, these "movers and shakers," as the news media dubbed them, had a prime concern in saving their investments in the city's central business district (Center City) by renewing and refurbishing it.
- The Reform Politicians, and the nascent Democratic Party. In 1945 Philadelphia, unlike other great cities of the Northeast and Midwest, was still ruled by the Republican Party, which had gained ascendancy in the post Civil War era. Thus the Democratic Party became the vehicle for political and governmental

> reform. The Reformers who came out of the local chapter of Americans for Democratic Action made allies with the established labor unions, black social, intellectual and religious leaders and pragmatic young politicians on the make. The politicians adopted the Reformers' program, which in turn was largely created by the Young Turks and the business establishment.

Together they created a peaceful revolution that changed the structure of the city. A cardinal tenet of that revolution was planning: physical, social and economic. In many ways Philadelphia was a pioneer in the field of urban planning. Ideas adopted here in the 1950's are still being hailed as new and innovative in cities across the nation.

One of the most successful public relations devices employed by the members of these three overlapping groups was the construction of a large scale detailed model of the city as it appeared in the late 1940's. The model featured revolving segments to demonstrate how the city could appear in the future. This model was on display for a number of years and had a great impact. It served to bolster the underlying assumption of these early planners that there are architectural solutions to social and economic problems.

In addition to producing a physical rendition of things to come, the Young Turks and the business establishment were able to prevail on the old Republican political leadership to create an Advisory City Planning Commission in 1949. This commission immediately began work on the construction of a comprehensive city plan. The ideas spawned in those early days are the ideas that still dominate city developmental programs.

To carry out these brave new plans the new forces essentially had to break the power of the old political machine, replace unskilled patronage employees with professional technocrats, and restructure the city government. Along with the Advisory City Planning Commission, a commission was set up to write a new city charter.

In 1951 the voters adopted this city charter and elected a Reform Democratic mayor, Joseph Clark.

The City Planning Commission was the centerpiece of this new charter. Also built into Philadelphia's constitution was a process of fiscal and physical planning. The charter mandated the creation of two annual budgets: an operating budget, which estimated the cost of running the city government and delineated the sources from which these funds would come; and a capital budget, which stated the cost of constructing new buildings and repairing old ones.

The charter required that the operating budget be balanced and forbade long term borrowing for its functioning. A ceiling was set by the state on the amount of long term money that could be borrowed to finance the capital budget. The charter also ordered the City Planning Commission to prepare the annual capital budget and to prepare capital budgets for the ensuing five years as well. Thus the charter insured that physical redevelopment of the city could be planned on a long term basis.

Then as now, Philadelphia's poor black population was seen as major cause of the city's problems and as an obstacle to economic rejuvenation.

In the 1940's when the city's intellectual planners began to grapple with these problems, the black population was heavily concentrated in the areas surrounding Center City. The planners originally sought to solve that problem by removing large numbers of blacks in the regions just north of Center City with an urban renewal plan, relocating them in a proposed low-income area nearly ten miles away at the outskirts of the city near the airport. Blacks south of Center City were to be pushed back and isolated by the construction of a very wide highway to be called the Crosstown Expressway.[5]

With the advent of the Reform Democratic mayors, who could not have been elected without the solid support of the burgeoning black population, such crude removal schemes were abandoned. But the objectives remained the same: to maintain middle and upper middle-class whites in the city.[6] This objective led to a heavy investment in renewing Center City and the construction of a whole complex of office towers known as Penn Center.

The city administration also arranged the underwriting with federal funds of a high-income residential area (Society Hill), the creation of Independence Mall to attract large company headquarters to areas previously occupied by small family owned business that were legally compelled to move, the rejuvenation of Market Street East, and the construction of a commuter rail system to unify the city's commuter rail services.

Nowhere is the class bias of these early planners more clearly revealed than in their transportation decisions. These decisions allocated a disproportionately large share of the transportation dollar to commuter rail services largely used by middle-income whites who constitute less than 20 percent of the total mass transit ridership. Buses and trolleys, used by the city's more numerous lower income residents, accounted for 80 percent of mass transit trips.[7]

During the reform years (1951–62) heavy emphasis was also placed on highways, which were again intended to benefit the relatively affluent car owner at the inadvertent expense of the mass transit rider. A plan was devised to construct a series of concentric ring highways around the central core of the city and the city itself with the thought that these roads would facilitate access to Center City and attract more workers and shoppers to help revive the lagging retail industry.

This same objective to hold and, if possible, increase the white population of the city, also led to the development of middle-class, suburban-style communities on open land in the Northeast and Roxborough. Eastwick, on the other hand, once planned as a holding territory for poor blacks, was newly conceived as a "city within a city." It was to be a model of a racially integrated, middle-income community.

Successful efforts were made to keep the universities within the city. To forestall suburban moves of major institutions, the city government encouraged and aided the expansion of Temple University, the University of Pennsylvania and Drexel University even when these expansions came at the expense of low-income neighborhoods.

Similar expansion encouragement was also offered to other city institutions such as Thomas Jefferson University and Hahnemann Medical College at the expense of retail and other taxable properties.

The years have taken a higher toll of the highway schemata than of any of the other concepts devised by these seminal planners. The last of their concepts, the development in Center City of Market Street East and the commuter tunnel, were finally under construction in 1981, more than three decades after they were originally conceived.

The early planners also conceived the Science Center located in West Philadelphia. Planned and developed by the city in collaboration with the University of Pennsylvania and Drexel University, its purpose is to act as a resource and magnet for businesses with high technology, scientific and/or intellectual needs.

Early in Mayor Clark's term (1951–55) the post of city economist was created. Its first occupant was Kirk Petshek, whose initial task was to begin a series of studies on manpower and employment.[8] These studies revealed what the mayor and the "movers and shakers" already knew: "that manufacturing jobs were declining both nationally and locally, that jobs were shifting into the service area, and that Philadelphia failed to participate in the national shift. . . . This shift plus the city's particular emphasis on manufacturing made the Philadelphia region the slowest growing in the nation in the fast-growing fifties."

To revise this grim picture, the city in 1954 began a concentrated effort to improve the employment situation. The first undertaking of the Clark Administration in the industrial field was the creation of a food distribution center in an effort to expand and modernize this industry.

A separate corporation was established. Bonds were floated. The city supplied construction capital and helped assemble the land. In 1969, fifteen years later, all the center's debts had been paid and the center employed 9,000 people.

From this initial success the city went on to devise a wide-ranging strategy whose prime objective was to lure existing business to Philadelphia.

A study was made to determine which industries would be most likely to respond to efforts to attract them to Philadelphia. Next an analysis was made to determine which firms or industries would be most desirable to have in the city. Mayor Clark established an Economic Advisory Committee to plan the city's economic campaign. This committee in turn decided on a set of criteria to be used to measure economic activities. The criteria were: 1) wage level and/or total payroll; 2) number of employees per acre; 3) economic stability; 4) kind of labor utilized; 5) likelihood of growth; 6) tax receipts; 7) linkage with suppliers and customers in the area; and 8) generation of secondary income and employment.

Significantly, as Kirk Petshek, who served on the Economic Advisory Committee admits,[9] "A criterion not considered at the time but which would be added today, is the question of how many hardcore unemployed could be absorbed by the industry, and whether the industry would be willing to participate in federal programs to assure their frictionless employment by the firm."[10]

In the decade between 1953 and 1963, $54,000,000 was spent on industrial renewal, or 24 percent of all the funds allocated by the city for urban renewal. Walter D'Alessio former Executive Director of the Philadelphia Industrial Development Corporation (PIDC), Philadelphia's main industrial renewal tool, views the city's entrance into comprehensive planning as a somewhat less straightforward action. "I would say Philadelphia tiptoed into the whole notion of comprehensive planning in 1956," D'Alessio recalls. "It was then that city officials realized that they had become the buyers of last resort of commercial and industrial real estate as industry and business moved out of the city or went under."

The problem of industrial flight, D'Alessio admits, was aggravated by the city's urban renewal and transportation policies. "It was then that we began to deal with the problem of government policy supposedly working for public good really hurting the public interest." The government was condemning land along the routes of the proposed Schuykill and Delaware Expressways and planning to create Independence Mall. These condemnations forced hundreds of small industries and businesses to move or close shop. "The city was losing jobs through its own policies. Our studies showed that we needed an antidote."

The antidote was PIDC. It was created in 1958 to provide aid to displaced business. It developed a series of industrial parks in the Northeast and Southwest and near North Philadelphia sections of the city. "The idea," D'Alessio explained, "was to provide alternative sites within the city to dislocated industries and thus save jobs." To facilitate these moves PIDC also set up a revolving fund in 1960 to supply low interest loans to its customers. A program of tax abatement was also instituted.

Since 1958 over a billion dollars in total projects have been assisted by PIDC; 1,300 transactions were conducted involving 142,450 jobs—63,200 of them newly created. The agency has sold over 1,000 acres, and 399 new buildings have been constructed, while 900 businesses have been relocated or expanded with PIDC assistance.[11]

This Philadelphia-bred concept of an urban agency which combines the functions of land bank, low interest money lender and tax abater has become one of the keystones of the newest national economic thinking. However, it wasn't until the mid 70's that the idea was taken up by such cities as Detroit, New York and Baltimore.

It is ironic that just as these ideas are receiving widespread acceptance questions are arising as to their efficacy. According to a study released by the Institute for the Study of Civic Values on Dec. 18, 1980, "the city (Philadelphia) loses $2.5 million annually in tax abatements to 145 companies on property assessed at $37,000,000." The Institute concluded that the tax abatement policy had no bearing on the number of jobs companies getting these benefits had in the city.

Questioned about these findings, D'Alessio said[12] his own partially completed study of the tax abatement program "shows that employment at thirty-four participating companies rose from 2,315 before abatement to 4,553 afterward. The question is: Would these deals be done without tax abatement? I don't know; it all depends on who's talking to them [the companies] and what they want to hear." Similarly it is not known how many recipients of PIDC low interest loans would have stayed in the city and/or expanded without this aid.

"In retrospect," D'Alessio commented, "PIDC works and it doesn't work. We have been able to keep some businesses in Philadelphia by providing land and financing. But we're not as successful as we had hoped to be even though we began to help a wide variety of companies besides those in danger from public condemnation. We failed because we could not keep manufacturing and blue-collar jobs here."

According to D'Alessio many of the city's key industries were doomed. "They either had to change the way they operated, or they had to go under." He pointed to the steel and garment industries and to the former Philadelphia Philco plants which manufactured TV sets. "Some of the garment industry moved south, and some moved to Taiwan and Hong Kong; but many of Philadelphia's garment factories simply went out of business as did much of the American steel industry. Philco did not relocate. It just shut its doors, and TV sets are not manufactured in the U.S. anymore."

D'Alessio pointed out, "There has been a new industrial revolution in the past quarter of a century. This revolution goes beyond the way we manufacture things to the way we do business in the United States. Whole categories of business have just disappeared."[13]

Sadder but wiser, city officials have more limited goals in the '80's than they did in the '50's. "We are trying to stabilize areas that are declining, but we know that in many cases we cannot stop the decline." D'Alessio took as an example the garment industry. "There is simply no way we can expect a revival of mass clothing production in Philadelphia because they have out of date plants, strict unions which call for such outmoded methods as hand cutting rather than laser beam cutters, and high wages. The textile industry has fled the Northeast and is now moving out of the country to Asia."

"However," the former PIDC official argued, "we can maintain and even help to expand that part of the industry which demands hand labor—the high fashion industry." In fact a number of high-style fashion companies now make their home in Philadelphia. They include Albert Nipon, the Dirrus Brothers, and Bleeker Street.

A second part of the city's current economic strategy is to recognize that there is room for growth in the health services industries. Philadelphia is a major medical center. It has five medical schools, more than twenty-five hospitals, and several major pharmaceutical companies. "It is logical for the city to become a center for the manufacture of medical appliances and related scientific equipment. We are working to encourage such development."

The current city administration is also attempting to make Philadelphia a center for the testing of drugs, and thus create another health related complex.

The thriving health industry includes the growth of nursing homes and retirement homes. "We believe that the nursing home industry and the retirement home industry can produce jobs, not just menial jobs but job ladders with interested workers being trained and moving up into the medical industry," D'Alessio added.

The city's economic planners are also looking to the expansion of the food processing industry and to the development of businesses created by the need for energy conservation. They also hope to develop industrial areas closer to the core of the city in land freed by abandonment.

It is envisioned that these new businesses will include fresh produce distributors; spin-offs from the chemical industry such as the manufacture of hard rubber plastics; kitchen products; subsidiary construction items, i.e., duct work, window frames, lighting fixtures; factories related to the electrical industry such as circuit breakers and coated wires; and mundane medical manufacturing such as test tube racks and casing for incubators. D'Alessio even sees Philadelphia as a growing site for the manufacture of minor munitions like shell casings, fuses and sighting devices.

According to D'Alessio, these enterprises will be looking for dependable labor with modest skills. "We believe they would be a perfect fit in North Philadelphia. The real issue now is who gets the jobs, and can we satisfy the job needs of the neighborhoods in which we try to place new industry and business.

"Power has moved out of the hands of the 'movers and shakers' and toward the advocates of lower-income communities. The emphasis on matching neighborhood economic development with neighborhood labor," D'Alessio admits, "reflects this change of political power. The city will invest in green lining and neighborhood industry. The 1980's in this vision of the future will be the decade of the neighborhoods."

Along with the need to renew the city physically and to provide economic aid to businesses in an effort to maintain and create jobs, city officials such as Mayor Clark and City Economist Petshek realized as early as 1954 that it was necessary to train workers for the new kind of employment that would be available. However, city funds were not allotted for this purpose, although city money was invested in both residential and industrial renewal.

Manpower training had to wait for monies from the federal government. They did not come until the Manpower and Development Training Act was passed in 1965 in response to ghetto riots in such cities as New York and Philadelphia. In 1970 this act was supplemented by the Comprehensive Employment and Training Act (CETA), which in 1981 will supply the city with $80,000,000 for manpower employment.

In the area of manpower training, as well as in the areas of reindustrialization, Philadelphia has played a leading role. The creation of Opportunities Industrialization Corporation (OIC) in the sixties by the Rev. Leon Sullivan has served as an international model for the training of unskilled workers.

In addition to producing federal funding for manpower training, the urban riots focused the attention of the city's political leaders on the appalling economic problems of poor blacks.

During the halcyon days of Philadelphia reform, Mayors Clark and Dilworth contented themselves with low overhead anti-discrimination programs which primarily benefited middle-class and upper-class blacks. These programs included the establishment in the city charter of a Human Relations Commission, the creation of a short-lived Police Advisory Board, and the appointment of prestigious blacks to important boards and commissions.

The riots also produced a new set of assertive black political leaders who began to make economic demands. An attempt to meet these demands was met by the federal government's creation of a panoply of urban projects collectively called the Poverty Program.

Locally the greatest single result of the 1964 black disturbances was the concentrated effort between 1965 and 1971 to upgrade the quality of public education. During this period there was a keen understanding that the traditional public schools had failed to educate large numbers of poor black children adequately. It was also recognized by the city's business, intellectual, and political leadership that there was a direct relationship between the quality of education of its work force and employment figures. Former Mayor Dilworth, who had been appointed to head a newly revamped Board of Education in 1965, stated that the way to save the city's economic base was to make it possible for the mass of poor blacks to qualify for skilled jobs.

To accomplish this end, Dilworth introduced a series of experimental educational programs into the system. Outstanding educators were recruited from across the country, old school buildings were replaced and new schools built in new communities. The largely segregated teaching and administrative staffs were integrated. A wide variety of plans was also developed to integrate the student body of the public schools, but none ever succeeded in accomplishing this goal.

In 1967 a large demonstration by black students was held in front of the Board of Education. The police broke up the demonstration and in the process produced a chaotic and disorderly dispersal of the crowd.

The demonstration proved to be a political turning point in the city's history. It dramatically revealed the sharp and angry cleavage between a majority of the white community and the black community. It stopped the Tate Administration's (1962–71) support of the Board of Education's special emphasis on innovative programs and poor black students. The demonstration also created a political platform for Police Commissioner Frank Rizzo's successful 1971 run for mayor.

Rizzo's elevation to office in 1972 signalled an end to special efforts to help the poor black community, and the cancellation of one set of elaborate plans for a bicentennial celebration that, it was hoped, would create thousands of jobs and houses for poor blacks. It also produced a new item on the city's economic planning agenda—neighborhood planning. The concept is eagerly embraced by the Green Administration.

During the Rizzo years planning was begun to revitalize

neighborhood shopping strips such as the American Street corridor. In this field, however, Philadelphia is a follower rather a leader. This national movement received its major impetus from the neighborhood movement in Chicago.

Although much of the neighborhood agenda was developed during the Rizzo years, and often by people and communities who saw this Mayor as their champion, the Rizzo Administration actually gave low priority to neighborhood economic development. Albert Gaudiosi, who served as Rizzo's City Representative and Director of Commerce, summed up that administration's philosophy in a recent interview: "No mayor ever had a greater affinity for the business community than Frank Rizzo. They got whatever they wanted. Rizzo was predisposed toward big business. He believed that the city had to have a viable Center City to maintain its tax base."

Consequently, the Rizzo Administration reemphasized the reformers commitment to Center City redevelopment by beginning to build three long-planned projects: the commuter tunnel, the first stage of retail store revitalization in the Market East area, and the conversion of Chestnut Street, an avenue of prestigious retail stores, into a pedestrian mall. The Rizzo years also saw the planning and construction of a pride of new central city office buildings and hotels. Gaudiosi explained:

> We had tremendous pressure from the unions on many of these projects. The unions wanted to get construction jobs for their members, and this was certainly a factor in setting our priorities. . . . I think that the tunnel will create jobs on the concourse level; and if we get Penny's on Market Street, the twin office towers and rehabilitation of Reading Terminal, we will have extended the city's tax base, created not only temporary

construction jobs, but also jobs in the new stores and facilities built because of the tunnel and Market Street East.

The Chestnut Street Transitway is another plan that had been on the books for a long time. It came to be built in the following way. Because of the relationship between Rizzo and Nixon, money was promised, six million outright to build the transitway in time for the Bicentennial. The Nixon administration was primarily interested in building an in-city shopping mall to see how it worked. Because of the Nixon-Rizzo relationship, we got first crack at being the model. We asked the Chestnut Street merchants if they wanted the transitway. They said yes. So we took the money.

So many of our decisions were based on where the money was; what the federal government would fund, we built.

But we certainly believed that top priority should go to the central business core in the hopes that it would prosper and pay more real estate taxes so that we wouldn't have to keep raising real estate or other taxes.

Essentially we reviewed ideas that were around. Some we discarded, such as the Crosstown Expressway; and now we have that wonderful renaissance on South Street where the highway would have been.

Philadelphia would be a great deal worse off in terms of employment now if we didn't have the tunnel going and the new construction of hotels and office buildings in Center City. Certainly that is why we are better off now than some other cities.

The truth is that each project was decided on its own and often by extraneous cross currents. But no mayor ever wanted to do more to please the business community or was more accessible to them.

We tried to give contracts to local firms; and if we couldn't, we tried to get outsiders to set up local Philadelphia offices.

We tried to get something from the state and federal governments in lieu of taxes.

We tried to work something out to stop the expanding of institutions with tax-free status from eating up taxable real

> estate. But with the drags those guys had, there was no way you could stop them.
>
> Rizzo's critics think that he had a fixation on Center City, but that is not true. We put together plans for neighborhood shopping strips and mini plazas.[14]
>
> Of course, everybody is talking about neighborhood rehabilitation now. But it is very, very difficult to do.

John Gruenstein, Chief of the Department of Research at the Federal Reserve Bank of Philadelphia and a man whose opinion is respected by the Green Administration, picked up Gaudiosi's theme:

> It is clear that Philadelphia neighborhoods need to be reinvigorated. But how do you do it? Should you put money into neighborhoods? Does it help? How do you motivate people? How can you help people? Will early failure make long term commitment untenable? Ultimately we could take the attitude that the city's economic life is kind of a sandbox, something you toy with and abandon since real life is taking place somewhere else.
>
> The truth is that very little is known about the shape of economic policies in the cities. We really don't know what are the forces that determine location. Those factors that ought to explain location really don't. In fact, nobody knows how to revise the decline of old manufacturing jobs. But service jobs are taking the place of manufacturing jobs. For too long we have put too strong an emphasis on manufacturing. There is, I believe, a myth about manufacturing. It is viewed as basic to productive work and that service jobs are ancillary. That is nonsense. We are a service-oriented economy. The shift is inevitable. We are at the kind of change-over that occurred in the nineteenth century when agriculture was the basic work and the vast majority of the people worked on the land. At that time manufacturing was viewed as ancillary. The industrial and agricultural revolution

changed all that. In advanced countries fewer and fewer people are needed to provide food. Agriculture has become a manufacturing industry. Now manufacturing is about to go the way of agriculture. With new technology fewer and fewer people will be needed to produce goods, and more people will be needed to give service.[15]

Gruenstein is encouraged by the fact that the City Planning Commission is contemplating looking at service jobs and creating a data bank on them. "Data exists on manufacturing jobs, but there really is no comprehensive data on service jobs," he states.

Despite the optimism of Gruenstein, D'Alessio, and Brown, the record of the past thirty-five years gives cautioning indicators. During these three and a half decades, Philadelphia did try to plan its economic destiny. It devised programs of tax incentives, land incentives, and technological assistance. It tried to keep businesses in the city, to lure industry from other areas, to help new firms get started and old firms expand, to improve the transportation network, to renew its physical plant, to reorganize the public schools, to upgrade the work force, and most recently to revitalize its neighborhoods.

Nonetheless, the job and industry drain continues. In the last quarter of 1980, the city saw the loss of another 1,000 jobs and the potential closing of four industrial facilities.[16]

The controversial Report of the President's Commission for a National Agenda in the '80's might well have been referring to Philadelphia's record of economic and urban planning when it stated, "governmental policies have failed to substantially relieve the economic hardships that afflict those who dwell in the nation's largest cities."

Notes

1. On October 16, 1980 the city of Philadelphia awarded $100,000 to develop a shopping mall at 9th and Columbia Avenue in cooperation with local merchants. The future mall is located in a low-income black neighborhood.

2. Campaign pledges of William Green 1979, collated by Deputy Mayor Phillip Goldsmith.

3. The Annual Planning Report for fiscal 1981, prepared by the Philadelphia Office of Employment and Security in May 1980, gave the following work force profile for the city of Philadelphia: blue collar, 18.8 percent; clerical, 29.2 percent; service, 36.1 percent; professional and executive, 15.9 percent.

4. The report placed Philadelphia's unemployment rate as overall 8.3 percent: white unemployment, 4.5 percent; black unemployment, 16.3 percent.

5. "Requiem for a Renaissance," *Philadelphia Magazine*, Nov. 1964.

6. Mayor Richardson Dilworth repeatedly talked about this problem during his six years in office. He invented the phrase "white noose" to describe the city's suburbs.

7. Bruce Caswell of the Institute for the Study of Civic Values.

8. Petshek's study was only the first in a long and continuing series of studies continuing into the present. The studies done in the late fifties and early sixties were the first to apply computer technology to social and economic problems.

9. Kirk R. Petshek, *The Challenge of Urban Reform* (Philadelphia: Temple University Press, 1973).

10. Ibid.

11. PIDC Annual Report, 1979.

12. *Philadelphia Inquirer*, Dec. 18, 1980.

13. D'Alessio cited the wholesale shoe jobbers, who are middlemen between manufacturers and retailers. "This industry used to be located in what is now the Independence Mall area. When we condemned the buildings which housed these businesses, we set aside space on Callowhill Street, one of the city's new industrial

parks, for them. Then we discovered that the shoe jobbers were mostly old men whose sons had gone on to other things. Rather than move, most simply went out of business. In fact the businesses have really disappeared. Shoe manufacturers no longer use middlemen. They deal directly with retail outlets."

14. These plans were made for American Street Corridor, in a working-class, white neighborhood, and Haddington Plaza and Allegheny West in working-class black neighborhoods.

15. The concept of America as a service economy has come under severe scrutiny. See *Newsweek* Magazine, Feb. 10, 1981.

16. *Philadelphia Bulletin*, Jan. 25, 1981 listed the following plant employment losses: 106, Abbotts Dairy, Inc.; 270, First Penna. Corp.; 200, General Electric; 300, DuPont Co.; 116, Budd Co.; and 126, Container Corporation.

Bibliography

Baltzell, E. Digby. *Puritan Boston and Quaker Philadelphia*. New York: Macmillan, Inc., The Free Press, 1979.

Brown, W. H., Jr. and Gilbert, C. E. *Planning Municipal Investment: A Case Study of Philadelphia*. Philadelphia: University of Pennsylvania Press, 1961.

Lowe, Jeanne R. *Cities in a Race with Time*. New York: Random House, 1967.

Petshek, Kirk R. *The Challenge of Urban Reform*. Philadelphia: Temple University Press, 1973.

Warner, William Bass, Jr., *The Private City: Philadelphia in Three Periods of Growth*. Philadelphia: University of Pennsylvania Press, 1973.

Weiler, Conrad. *Philadelphia: Neighborhood, Authority and the Urban Crisis*. New York: Prager Publishers, Inc., 1974.

Interviews

Binzen, Peter. *Philadelphia Bulletin* Business Reporter
Brown, Allan L. Deputy Director of Commerce

D'Alessio, M. Walter. Executive Vice President, PIDC

Gaudiosi, Albert. former City Representative and Director of Commerce

Gruenstein, John. Research Officer and economist, Federal Reserve Bank of Philadelphia

Schwartz, Edward. President, Institute for the Study of Civic Values

Summers, Anita. Adjunct Professor of Legal Studies and Public Management, and Associate Chairman for Public Management, Wharton School, University of Pennsylvania

Barbara Baker: Campaign for Human Development, U.S. Catholic Conference—used with permission

Delaware Valley Coalition for Jobs

Top: John Dodds, coordinator of the Delaware Valley Coalition for Jobs, organizing the campaign against a plant closing by Container Corporation of America in Manayunk. *Bottom*: workers at the CCA plant forming up for a demonstration, December 4, 1980.

Chapter 8

Capital Flight and Job Loss: A Statistical Analysis

Arthur Hochner and Daniel M. Zibman

In the greater Philadelphia Standard Metropolitan Statistical Area (SMSA), there was a net loss of over 126,000 manufacturing jobs in the decade 1969–1979.[1] In a review of the Philadelphia economy's prospects for the 1980's, three local business journalists concluded: "Most of the manufacturing jobs lost here in recent years have simply disappeared. The factory owners did not pack up for the Sunbelt states or the West Coast. Companies, for the most part, simply went out of business because it costs too much to operate in Philadelphia." Businessmen were quoted in the same article: "We've put a lot of our problems behind us," said Henry Wendt, president of SmithKline Corp. "The weak industries have closed or been reduced. . . . I believe that we've shaken out most of the marginal industries," said James O'Brien, manager of area development for Philadelphia Electric Co.

More recently, the new president of the Philadelphia Federal Reserve Bank was quoted: "I think there is something to the Darwinian theory," Edward G. Boehne said. "Any company that survived the 1974–75 recession," he said, "must have been strong."[2] Thus, what is happening, according to this view, is unfortunate but necessary in the Darwinian struggles of the free market economy.

However, there is another explanation of the plant-closing/job-loss phenomenon. It can be called the capital shift model. This model emphasizes the dynamic power of capital mobility.[3] This mobility has taken a quantitative leap with the growth of great conglomerate and multinational corporations. Capital shifts are present in any dynamic economy. However, the nature of huge conglomerate and multinational enterprises has enabled the process to be accelerated. Without loyalty to any particular industry, region, or country, these corporations are free to invest their money in the most profitable and fastest growing sectors or locations. It is the constant search for growth that accelerates capital movement and, consequently, increases the frequency of plant shutdowns.

In other words, the capital shift model posits that the plant-closing/job-loss phenomenon does not occur primarily as the result of inexorable, natural market forces and the decisions of millions of individuals. Rather it states that the decisions of a rather small group, the executives of large multinational and conglomerate firms seeking high corporate growth rates, cause severe economic dislocations. Furthermore, the decisions are the outcome not of the free market but of economic and political concentration which gives these enterprises great power and of federal and state government policies on taxes, subsidies, loans, defense spending, labor law, and economic development which favor the multinational and conglomerate firms.

The plant-closing/job-loss phenomenon has always been a concern. But concern has increased greatly in the 1970's, coincident with the great conglomerate merger movement of the late 1960's, when mergers averaged 3,500 per year. As Mark Nadel notes, "By 1968, 89 percent of all mergers were conglomerate in nature."[4] And in the latter 1970's, the merger movement picked up steam again. Moreover, the late 1960's and early '70's coincided with enormous growth of foreign investment by multinational corporations.

Absentee ownership, of which multinational and conglomerate ownership is a part, has traditionally been associated with such phenomena as a lack of employment stability, lower wages, reduced civic involvement by business, and lower contributions to charitable agencies. Studies of the effects of absentee ownership, especially by conglomerates, point further to the deleterious impact of conglomerate acquisition of local firms on the local establishments' employment and sales growth rates. Conglomerates tend to acquire or merge with firms that are winners—with high growth rates—and then on average slow their growth rates to less than the average of nonacquired firms.[5]

The objective of overall corporate growth rather than mere profit-making by a division/subsidiary is the key to understanding the dynamics of huge corporations and the effects of absentee ownership. Growth requires continual investment of capital. Conglomerates and multinationals need their divisions and subsidiaries to be highly profitable to provide much of the capital. Thus, they target certain rates of return for these corporate arms and failure to reach the target leads to divestiture or shutdown. Furthermore, they acquire cash-rich and profitable enterprises to use as "cash cows" for funding other investments. The upshot is that even profitable corporate subdivisions are shut down if they are not achieving target profits or if their profits and cash flows are milked for use elsewhere in the corporation.

Multinational investments may accelerate corporate growth by bringing competitive advantages—cheap labor markets, closeness to supplies or customers, little regulation, and weak labor unions. These conditions are best achieved outside of the U.S. Thus, multinationals can and often do shift capital away from the U.S. and into growth elsewhere. The extent to which foreign investment by U.S. multinationals has reduced or increased domestic employment has been strongly debated for the past decade. A recent econometric study, however, estimates that from 1966 to 1976, the U.S. lost 1.06 million jobs as a result of direct foreign investment.[6]

In sum, the capital shift model considers plant closings and resultant job loss only the most visible signs of the economic dislocations caused by private decisions concerning capital investment and corporate growth. Because capital shift is so private and relatively invisible to as a cause of economic dislocation, statistics are hard to come by. However, some links between capital shift and job loss can be found in statistics concerning the Philadelphia metropolitan area.

A study of the closing of 173 firms in the Philadelphia SMSA between 1976–79 reveals the important role played by conglomerates and multinationals.[7] The study covered 61,011 manufacturing jobs, or 48.4 percent of the 126,000 jobs lost in the past decade.

Table 8-1 shows the distribution of establishments shut down or relocated among the three ownership types of interest. In the overall sample, and in the manufacturing sector in particular, where the largest number of closings was recorded, the IND's accounted for the largest single proportion (42.8 percent and 46.4 percent respectively). However, together the AOC and MNC/CONG categories accounted

Table 8-1. Establishments Shut Down or Relocated in Philadelphia SMSA, by Ownership Type and Major Industry 1970–1979*

	Ownership Category			
	IND†	*AOC*‡	*MNC/CONG*§	*Total N*
Overall Sample	42.8%	24.9%	29.5%	173
Manufacturing (SIC #s 20–39) ‖	46.4%	23.9%	29.7%	138
Retail trade (SIC #s 52–59)	31.3%	43.8%	25.0%	16
Wholesale trade (SIC #s 50–51)	16.7%	16.7%	66.7%	6
Federal govt. (SIC # 91)	—	—	—	5

*Major industries with fewer than five establishments shut down or relocated were omitted from this table.

†IND = Independently owned

‡AOC = Absentee-owned companies (from outside Philadelphia SMSA)

§MNC/CONG = Multinational or conglomerate corporations

‖ SIC stands for Standard Industrial Classification codes.

for 54.4 percent of the overall total and 53.6 percent of the manufacturing total. It is to be expected that IND's, being smaller and, we assume, economically weaker establishments, would account for a very large proportion of shutdowns (see Table 8-2). However, what is surprising here is that over half of the shutdown firms came from the presumably more mobile sectors, i.e., those with ability to shift capital rapidly and likely to be more adaptive in a Darwinian sense. The remainder of the closings are attributable to the

Table 8-2. Mean Size of Establishments Shut Down or Relocated in Philadelphia SMSA, by Ownership Type and Major Industry 1970–79

	Ownership Category			
	IND	*AOC*	*MNC/CONG*	*Over-all Types*
Overall sample	286	694	473	486
Manufacturing	308	629	501	442
Retail trade	183	1,250	456	718
Wholesale trade	250	100	324	274
Federal govt.	—	—	—	1,780

federal government (perhaps the largest multinational/conglomerate operation in the world). These government closings were also the largest, averaging 1780 jobs each.

Table 8-3 shows the distribution of jobs lost among the ownership types. Overall, and in every category, IND's accounted for no more than one third of the jobs lost. Together, the AOC and MNC/CONG categories accounted for at least three fifths of the jobs lost and up to ninety percent in some industries. That is, if these assumptions are correct, firms with the greatest capital, not the weakest and most marginal, were accountable for the greatest job loss. Moreover, in the manufacturing sector, the MNC/CONG category alone covered over one third of the jobs lost.

A surprising result is the very low proportion of jobs lost among IND's in the retail and wholesale trade industries compared to the proportions from the AOC's and MNC/CONG's. In these industries, we would expect the IND's, i.e., the weakest firms, to be squeezed out as population drained away from the region, and for the AOC's and MNC/

Table 8-3. Jobs Shut Down or Relocated in Philadelphia SMSA by Establishment Ownership Type and Major Industry 1970–79

	Ownership Category			
	IND	*AOC*	*MNC/ CONG*	*Total N*
Overall sample	25.2%	35.5%	28.7%	84,054
Manufacturing	32.3%	34.0%	33.7%	61,011
Retail trade	8.0%	76.2%	15.9%	11,493
Wholesale trade	15.2%	6.1%	78.7%	1,645
Federal govt.	—	—	—	8,900

CONG's to be able to resist the pruning of department and food stores, owing to their greater resources. However, here too, as in manufacturing, the most mobile firms do the greatest amount of disinvestment.

Some relatively sketchy data we found add some interesting information about the continued fortunes of the firms in the sample. Among the seventy-four IND's, we found such data on only five firms. Among this group employment figures were steady, and the median increase in sales was 43 percent over ten years. Obviously, they failed to keep pace with inflation.

Among the 43 AOC's, data were discovered concerning seventeen firms. The median increase in employment was 9 percent and the median increase in sales was 122 percent. That is, on average, AOC's which shut down operations in the Philadelphia SMSA kept up with inflation and slightly increased their employment.

Among the 51 MNC/CONG's, clear data were discovered on thirty firms, the highest percentage. Here the median increase in employment over the ten-year period was zero.

However, the median increase in sales was 133 percent, better than inflation. Furthermore, of the 22 MNC/CONG firms increasing in sales by over 100 percent in the ten-year period, eight reduced employment. In other words, the most competitive firms tend to continue to do good business even in an inflationary economy, but do not have any overall beneficial impact on employment. They may be able to move their resources to the places of greatest profitability and competitiveness, but the capital shift process does not seem to create many jobs.

It might be argued that manufacturing is in decline generally, and that, since each ownership category has about the same percentage of the total job loss, the process that affects all three is the same; namely, an unprofitable location and declining product demand. We cannot rule out such an explanation. Unfortunately, we do not have enough data about the workings of the three types of firms in regard to job creation and other matters to fully examine that argument.

However, not every manufacturing industry is the same in the relative impact of the three ownership types on job loss. The relative impact varies by industry (see Table 8-4); that is, in printing and publishing, IND's acccounted for one half of the jobs lost; but in electric and electronic equipment, they accounted for less than one tenth; and in transportation equipment, for none. Nevertheless, in only one of these seven major industries was the IND sector responsible for half or more of the jobs lost. Over the seven industries as a whole (covering 45,050 jobs lost), IND's accounted for only 29.5 percent of the total. MNC/CONG's alone, on the other hand, accounted for over one half of the jobs lost in three of these industries and for 37.4 percent of the jobs lost in all seven industries. Even in the garment industry, where many small IND firms folded, the MNC/CONG's accounted for over half of the job loss. Where the greatest manufacturing job loss

Table 8-4. Jobs Shut Down or Relocated in Philadelphia SMSA by Establishment Ownership Type 1970–79*

	Ownership Category			
SIC#†	*IND*	*AOC*	*MNC/ CONG*	*Total N*
20. Food & kindred products				
Establishments	27.3%	27.3%	45.5%	11
Jobs lost	21.9%	43.6%	34.5%	4,039
23. Apparel				
Establishments	75.0%	10.0%	15.0%	20
Jobs lost	40.5%	8.8%	50.6%	7,350
27. Printing & publishing				
Establishments	53.8%	38.5%	7.7%	13
Jobs lost	50.6%	33.4%	16.0%	5,690
33. Primary metal ind.				
Establishments	45.5%	27.3%	27.3%	11
Jobs lost	38.1%	44.5%	17.4%	10,780
35. Machinery, except elec.				
Establishments	56.3%	12.5%	31.3%	16
Jobs lost	44.2%	7.6%	48.2%	4,040
36. Electrical & electronic equipment				
Establishments	16.7%	16.7%	66.7%	12
Jobs lost	9.5%	39.5%	51.0%	8,176
37. Transportation equip.				
Establishments	0	20.0%	80.0%	5
Jobs lost	0	40.2%	59.8%	4,975

*This table includes only those industries with shutdown or relocation of over 4,000 jobs in the ten-year period.

†Standard Industrial Classification code.

occurred, MNC/CONG's were generally responsible for the largest proportion. To reiterate, the most mobile, largest, and most apparently successful firms (MNC/CONG's and AOC's) were responsible for the overwhelming majority of jobs lost among the 173 firms from the Philadelphia SMSA in the sample over the seventies.

In general, the findings of this study offer firm support for the capital shift model's explanation of the plant-closing/job-loss phenomenon, at least in the Philadelphia area. Furthermore, this study's findings show the key role played in job loss by absentee control of several kinds, including domestic corporations, multinationals, and conglomerates.

In light of these findings, the conventional wisdom about plant closing, especially the version circulating in Philadelphia, is shown to be off the mark. It is not simply the marginal, inefficient, and unprofitable firms going out of business in some survival-of-the-fittest context.

But perhaps that could have been known even without this study. Philadelphia commentators on the problem often point out that of the 126,100 manufacturing jobs lost in the decade 1969–79, in the year 1974–75 alone about 51,600 manufacturing jobs were lost in that sector. Thus, they conclude, the recession of that period cut out the weak and marginal firms. However, looking at the changes in manufacturing employment for earlier years in the decade, one finds a somewhat different picture.[8]

From 1969–72, there was a net loss of 77,700 manufacturing jobs in the Philadelphia SMSA. That is, manufacturing employment was dropping drastically much earlier than the 1974–75 recession, when manufacturing job loss reached its peak. The recession was important, but by no means the entire story. Perhaps there is more than a coincidence between the high job loss from 1969–72 and the giant wave of conglomerate mergers in the late 1960's. The merger movement has had the effect of dropping the Philadelphia SMSA from number three as a corporate headquarters area in 1967 down to a tie for seventh place in 1980. Numerous jobs were lost in the SMSA as a result of the mergers.

The post-industrial transformation of American society and the demographic deconcentration of the Frostbelt cities may indeed be happening, but the underlying dynamics may have more to do with corporate balance sheets, concentration of wealth and power, and increasingly rapid capital mobility. Philadelphians lost their jobs not so much because of their region's failure to adapt to natural changes as because of the remarkable quickness of multinationals and conglomerates to adapt themselves to ways of improving their growth pictures. The capital shift model and the findings supporting it here have many implications for public policy. Legislation controlling plant closings, conglomerate merger investigations, tax policies on multinationals and differing tax structures in the various states, and labor law reforms may all need to be considered in devising a program to combat unemployment and economic dislocation. Although it was not possible in the scope of this study to look at the presence or absence of unions at the establishments which closed or relocated, the findings have several implications for the labor movement. Rapid and unrestricted capital mobility not only threatens jobs, but also the viability of unions in the heavily unionized Frostbelt. The problem is not merely runaway factories, but, more broadly, runaway capital. Unions may have to be more aggressive, not only in organizing workers in the Sunbelt, but also in pushing for bargaining over disinvestment decisions. Multinational and coalition bargaining may take on more importance to deal with the policies and practices of the multinationals and conglomerates. Furthermore, increasing union control over pension fund investments may increasingly be used to counter the corporate trends towards rapid disinvestment.

Notes

1. U.S. Department of Labor, U.S. Bureau of Statistics, *Philadelphia Employment Trends*, June 1980. The 1980 census places this figure at 140,000, which is the figure now accepted by the City of Philadelphia and used in other chapters of this book.

2. Douglas A. Campbell, "Philadelphia into the 1980's," *Philadelphia Inquirer*, Jan. 7, 1980, p. 4D; Robert L. Waters, "Plant Closings Fire Debate on Business Ethics," *Philadelphia Inquirer*, Feb. 8, 1981, pp. 1–2D; Andrea Knox, "Manufacturing in Philadelphia Seen on Firmer Ground," *Philadelphia Inquirer*, March 11, 1981, p. 7D.

3. See, for example, Barry Bluestone and Bennett Harrison, *Capital and Communities: The Causes and Consequences of Private Disinvestment* (Wash. D.C.: Progressive Alliance, 1980).

4. Mark V. Nadel, *Corporations and Political Accountability* (Lexington, Mass.: Heath 1976), p. 122.

5. See *Conglomerate Mergers—Their Effects on Small Businesses and Local Communities*, Hearings before the Subcommittee on Antitrust and Restraint of Trade Activities Affecting Small Businesses, Committee on Small Businesses, House of Representatives, 96th Congress, 2nd session (Wash. D.C.: Government Printing Office, 1980). Of special interest in this volume are the following reprinted statements and papers:

- Robert N. Stern and Howard Aldrich, "The Effect of Absentee Firm Control on Local Community Welfare: A Survey"
- Jon G. Udell, "The Community and Mergers: Social and Economic Consequences of Corporate Acquisition on a State"
- David L. Berkley, "Plant Ownership Characteristics and the Locational Stability of Rural Iowa Manufacturers"

- David L. Birch, "The Job Generation Process"
- William Norris, "Irresponsible Mergers and Acquisitions."

Although most of these statements come from academics, it is worth noting that Mr. Norris is chairman and chief executive officer of Control Data Corporation, 159th in *Fortune*'s 1980 listing of the 500 largest industrial corporations.

6. Robert H. Frank and Richard T. Freeman, "Multinational Corporations and Domestic Employment," paper for the Dept. of Economics, Cornell University, 1977.

7. For a complete exploration of the study and its methodology see Arthur Hochner and Daniel M. Zibman, "Plant Closings and Job Loss in Philadelphia: The Role of Multinationals and Absentee Control," presented to the 1981 Annual Meeting of the American Sociological Association, Toronto, Aug. 1981. Original data came from a report issued by the Commonwealth of Pennsylvania.

8. U.S. Department of Labor, U.S. Bureau of Labor Statistics, *Philadelphia Employment Structure and Trends, 1975* (Wash. D.C.: Government Printing Office, 1976); U.S. Dept. of Labor, U.S. Bureau of Labor Statistics, *Philadelphia Employment Trends, 1979* (Wash. D.C.: Government Printing Office, 1980).

Chapter 9

Clothing and Textiles: The Departure of an Industry

Pamela Haines

In 1870, Philadelphia, the major industrial city of the nation, led the country in textile production, and was the greatest center of woolens, hosiery, and textile machinery. The clothing industry had received major impetus from the demand for uniforms during the Civil War and was off to a booming start.[1] On the eve of World War II, although textiles had started a significant shift to the South, the city was still a giant in both industries. It was one of the leading producers of cotton fabrics, home furnishings, uniforms, shirt collars and night wear, cotton yarn and thread; and it was also an important national producer in rugs and carpet, upholstery, men's and boys' clothing, worsted goods, women's clothing, and dyeing and finishing. The two industries together used one-quarter of Philadelphia's industrial plants and employed one-third of its workers; and the clothing and textile workers' unions were among the best-organized, strongest, and most respected in the city.[2]

In 1950, there were 42,000 textile workers in the five-county Philadelphia area, and clothing worker union membership stood around 20,000.[3] By 1979, however, only about 1,800 textile workers and fewer than 8,000 clothing workers remained in the Philadelphia union. The industry had moved south and overseas.[4]

Although the clothing and textile industries of Philadelphia emerged from World War II in a strong position, the underlying industrial trends were ominous. Local employment in textiles had dropped by a third between 1914 and 1945 and continued steadily downward after World War II. Changes in style and consumers' habits, growth of synthetics, and advances in technology all contributed to making old plants obsolete; and union wage rates led to competitive disadvantage.

In the clothing industry the situation was uniformly less gloomy from the beginning, but the contributing factors and final outcome were similar. After World War II there was a boom in men's clothing production, both to meet consumer demand, pent up during the war years, and to clothe returning soldiers. Along with the boom came changes in styles. The new casual suburban lifestyle brought with it a shift in the industry from almost exclusive production of suits to greater emphasis on sportcoats and separate pants. New fabrics also came in: whereas men's clothing (with the exception of cotton shirts) had traditionally been wool, synthetics were rapidly entering the market. As the years passed, both these trends continued. Informal men's wear assumed increasing importance as jeans joined sportcoats and synthetics steadily replaced cotton and wool.[5]

These fashion changes had direct impact on production location. There was no necessary link between the new lines of production and the traditional centers of the men's clothing industry: the "suit towns" of the industrial North (New York, Philadelphia, Rochester, Cincinnati, Chicago). The industry had always been a mobile one, with low demand for fixed-site raw materials, and minimal fixed plant and machine expenditure. New firms could locate almost where they pleased, but a number of factors influenced their location decisions. With space at a premium in the old industrial cities, site costs tended to be high. Other disamenities to these old urban areas were assuming ever greater significance. Labor cost also continued to be an important issue in such a labor-intensive industry, where labor averaged 35 percent of total costs as compared to the norm of 5–7 percent. Especially with the passage in 1947 of the Taft-Hartley Act, which legitimized state open-shop legislation, employer after employer decided on the South or Southwest as more attractive places to locate in than in the unionized and high-wage Northeast. In 1947, the South accounted for roughly a quarter of the production in the non-suit segments of the United States apparel industry; by 1972 the figures were close to three quarters.[6]

The "suit" towns continued to produce and to grow from some of the new demand, but the most rapidly expanding sector of the industry in the 1950's and 1960's was in the South and Southwest. As new synthetics-dominated firms began to expand their line of production to include suits, they cut even more into the old centers' share of the trade. The suit industry, which had been almost completely organized in 1938, decreased to about 80 percent in the late 1970's. By the same time, the cotton garment industry, never so strongly unionized and more affected by new non-union growth, had close to 25 percent organization.[7]

Despite all these changes, the high level of demand generally kept the entire men's clothing industry of the country healthy through most of the 1960's. Though the Philadelphia men's clothing industry did not grow as phenomenally as it might have with as large a share of the newer lines of production as of suits, both the industry and the union membership steadily increased. There was a general sense of prosperity and complacency.[8]

The second and decisive blow came toward the end of the 1960's when imports began entering the U.S. market in rapidly expanding volume. The same factors that induced apparel and textile manufacturers to head south also lured them to underdeveloped countries abroad. The need for sophisticated technology or major capital investment was low; and the key ingredient, cheap labor, was abundant. (A striking confirmation of the importance of the latter, paralleling the United States industry's move south, appears in the shift of the Japanese clothing industry's production to Korea after the Japanese workers organized in the 1970's.) The labor cost differential is significant. In 1972, for example, clothing industry labor costs averaged $3.26 per hour in the United States, 94.3¢ in Japan, 44.4¢ in Hong Kong, 17.4¢ in Taiwan, 16.3¢ in Korea, and 12.5¢ in Columbia.[9]

Both indigenous foreign industry and subsidiaries of U.S. multinationals have found clothing a profitable product for export. Loopholes in the U.S. tariff laws allow manufacturers to send out parts to plants in low-wage countries for "assembly," then bring them back in almost duty-free. The production costs of imports in general are much lower than those of American-made clothing. On the other hand, while occasionally dipping with the entry of new imported items, prices go right back up to the level of comparable American-made products as soon as the market is secure.[10]

The impact of these imports on the U.S. clothing industry has been overwhelming. In 1965 the wardrobe of the average American male was almost 100 percent U.S.-made. By 1976 30 percent of the shirts, 30 percent of the sportcoats, 18 percent of the trousers, and 12 percent of the suits were made

Gail S. Rebhan

Dick Rothman: Amalgamated News

Dick Rothman: Amalgamated News

Amalgamated Clothing and Textile Workers Union carries the J. P. Stevens Boycott Campaign to the streets of Philadelphia and Washington, D.C.

abroad. In the same period, employment in the U.S. clothing industry declined by 25 percent while overall employment in the country rose by 32 percent. Between 1973 and 1975 alone, the U.S. apparel industry lost 65,000 jobs; there is no indication that the trends are changing.[11]

In addition to changes in style, the shift of the industry toward the South, and the massive influx of imports, a fourth factor of national and international scope with major impact on the clothing industry has been the growth of conglomerates. Traditionally the industry has been one of small shops, family-owned companies, and partnerships. Even in the late 1970's the Amalgamated in Philadelphia had contracts with over three hundred employers in the city. Often the company was started by an ex-clothing worker himself. Furthermore many employers still participated in some part of the production process. Although there are significant differences between management and labor even in such a personalized setting, there is an important sense of common interest in the clothing industry as a whole. For both employer and employee, the immediate livelihood depends on the company's maintaining enough of a share of the industry to keep marketing its goods. The small companies, too, have been threatened by imports as much as the workers, and have struggled hard to stay afloat.[12]

Since the late 1960's, however, conglomerates have entered the clothing industry for the first time in significant numbers. With a much larger financial base and a greater variety of holdings, their survival is not so easily threatened by the loss of any one plant, or even any one industry. They can afford, therefore, to be purely profit-oriented. If they buy up a small clothing manufacturer and discover that the profit margin is not so high as could be elsewhere, they are quick to close it down and write it off. For example, in Philadelphia, the Louis Goldsmith Company, employing four to five hundred workers, was purchased by Kaiser-Roth, which in turn was bought by Gulf and Western. Soon thereafter the plant was shut down completely. While this trend is certainly not limited to apparel manufacturing, the combination of the growth of conglomerates and that of imports—not unrelated as often the conglomerates own the overseas operations—has been devastating to the unionized clothing industry in this country.[13]

The changes since World War II in the Amalgamated Clothing Workers Union in Philadelphia, in numbers of members, type of work, and general strength, make sense in light of these larger trends. The union came out of the war with its pre-war strength of roughly twenty thousand. While style changes caused a decrease in suit production, that was more than made up for by rapid growth in other areas, especially cotton and washable garments. By the end of the 1960's, the union claimed a membership of 25,000. This was the time of greatest prosperity. There was more work than workers. Ads for employees appeared in three languages in the newsletter. Training programs for operators were established. Each month the union manager's report sounded better than the last.[14]

Then, starting in 1969, the situation reversed. With the increase in imports, local trade declined steadily. Shop after shop closed its doors. Operators were working only two or three days a week. The union was on the defensive, battling no longer to hold its own but just to minimize its losses. Each month's report was gloomier than the last's. The issue was not how to train new operators, but how to get the best deal in severance pay. The closing of the Daroff plant, holder of the prestigious Botany 500 label and employer of over 1,100 workers, consumed the attention of the union staff all through 1972 and 1973. Union membership hit an all-time low.[15]

There is some indication that the trend will not continue until the industry has entirely disappeared from the city. Several of the large firms that remain seem to have carved out a fairly secure niche, where, in their production of high-quality wool suits and topcoats, the demand will perhaps remain low but constant. But it is highly unlikely that the union will ever regain its former strength in the city. As the education director puts it, "There's just nobody left out there to organize."[16]

Notes

1. Gladys L. Palmer, *Philadelphia Workers in a Changing Economy* (Philadelphia: University of Pennsylvania Press, 1956), pp. 8–9; Sam Bass Warner, Jr., *The Private City: Philadelphia in Three Periods of Its Growth* (Philadelphia: University of Pennsylvania Press, 1968), pp. 69–71.

2. Federal Writers' Project, WPA, *Philadelphia: A Guide to the Nation's Birthplace* (Philadelphia: William Penn Association of Philadelphia, 1937), pp. 147–48; Wilfred Jordan, *Philadelphia: A Story of Progress*, vol. III (New York: Lewis Historical Publishing Co., 1941), pp. 73–74.

3. Palmer, *Philadelphia Workers*, p. 35.

4. Bernie Dinkin, Education Director, Philadelphia Joint Board, Amalgamated Clothing and Textile Workers Union, interviews, summer 1979.

5. Sara Fredgant, retired Education Director, Philadelphia Joint Board, ACTWU, interview, summer 1979.

6. Fredgant interview; Dinkin interview; ACTWU, *Summer School Manual* (New York: ACTWU Education Department, 1979), n.p.

7. Dinkin interview.

8. Ibid.

9. ACTWU, *Summer School Manual*, sect. I, "Imports."

10. Ibid.

11. Ibid.

12. Dinkin interview.

13. Ibid.

14. Philadelphia Joint Board, ACTWU (name changed from Amalgamated Clothing Workers of America in 1976), *Amalgamated News*, Oct. 1966, Dec. 1965. The tone of prosperity can be sensed in any issue of the newsletter between 1965 and 1968.

15. Dinkin interview. The tone, again, can be sensed in any issue of the *Amalgamated News* from 1969 to 1979.

16. Dinkin interview.

Chapter 10

Tasty Baking Company: The Company That Stayed

Lynne Kotranski and Douglas Porpora

Most discussions of capital disinvestment focus on companies that have left an area and the precipitating reasons and consequences of such a move. It can be just as important, however, to focus on a company that entertained the possibility of moving and decided against it. The Tasty Baking Company of Philadelphia is such a company. This account will consist of four parts: first, a brief history of the company and its present structure/organization; second, an elaboration of the reasons that led Tasty to consider moving from its present location; third, the circumstances involving the final decision not to move; and fourth, the organizational efforts to offset the conditions that suggested a decision to move in the first place and the consequences of these efforts for the old neighborhood.

The Tasty Baking Company was founded in Philadelphia in 1914 by Philip J. Baur and Herbert C. Morris. Their main product is the prewrapped, single-portion Tastykake. The Company, originally located on Sedgely Avenue, has been located since 1923 at 2801 Hunting Park Avenue in the "Allegheny West" area of North Philadelphia.

Up until 1961 Tasty Baking Co. was a privately owned, "family" run business. In 1961 Tasty became a publicly held company, but major stock holdings were still in the hands of family members. In 1965 Tasty began to diversify. In addition to its Philadelphia bakery operations, Tasty acquired, (1) Phillips and Jacobs, Inc., a graphic arts distribution firm; (2) Buckeye Biscuit Co., an Ohio distributor of cookies; (3) Larami Corp., a national toy distributor based in Philadelphia; and (4) Ole South, a frozen food operation. The latter was sold in April, 1979 and represented a loss of $6.7 million to the company. It was the first such loss ever in the company's sixty-six year history.

Even after diversification, the Tasty Baking Co. cannot be considered a very large company by national standards. However, in 1968, just about the time that Tasty was considering moving out of Philadepphia, it was the eighth largest industrial firm in Philadelphia. In 1979 the company had net sales of $169.5 million, but took an income loss of $2.3 million. The highest profit yielding year within the last decade was 1976, when net income totalled $6.7 million. In comparison with other large corporations, such as Du Pont Co., which made a profit of $939 million in 1979, the Tasty Baking Co. could almost be considered a "mom and pop" operation. Nevertheless, the Hunting Park plant employs a sizable work force—1,800 of the company's 2,100 employees.

In early 1968 Tasty Baking Co. began to consider the possibility of moving from its North Philadelphia location. One condition motivating such a consideration was the socioeconomic and demographic changes that characterized the Allegheny West neighborhood.

Prior to 1960 Allegheny West was a stable, white, lower-middle-class community, composed of several different ethnic groups (Italian, Irish, Jewish). However, between 1960 and 1970 Allegheny West changed radically:

- the racial composition of the area went from 96 percent white in 1960 to 86 percent nonwhite in 1970;
- owner-occupancy decreased from 83 percent in 1960 to 74 percent in 1970, although it was above the city's average of 60 percent.
- the unemployment rate jumped from 5.0 percent in 1960 to 7.2 percent in 1970;
- in 1970, the percent of families below the poverty level (16 percent) and those receiving public assistance (40 percent) were above the average for the city;
- the racial composition of schools changed drastically; 94 percent of persons nineteen 19 years of age or less were nonwhite;
- housing became overcrowded because of increases in family size of immigrating residents;
- median years of school completed fell below high school level;
- the occupational status of neighborhood residents changed: in 1960, 33 percent of the population were employed in white collar or professional jobs; by 1970 this figure was 23 percent. The percentage of unskilled workers increased from 18 percent to 32 percent over the decade:
- concomitant with the above changes, property and personal crime was increasing. Small businesses and industry, plagued by incidences of vandalism and high security expenses, were moving out of the area (eg., Link Belt, Nicetown Ball Bearing, Sears Warehouse, etc.).

Finding itself in a neighborhood undergoing transition left Tasty Baking Company with three options: (1) the company could move; (2) the company could stay without attempting to intervene or reduce the negative effects of rapid change (simply hire guards, build a fence around the plant, etc); or

(3) the company could stay in Allegheny West but take some steps to ameliorate the conditions plaguing the community. Tasty decided to exercise the third option.

In 1968 Tasty Baking Company made the decision not to move from Allegheny West. The motives were partially due to an act of social conscience and ethical ideals not to abandon the "old neighborhood." However, the primary reason not to move was based on hard-nosed economic considerations. Of these reasons, perhaps the most important was Tasty's satisfaction with and investment in its present plant.

By 1968 Tasty Baking Company had invested a great deal of time and money in improving and modernizing its Hunting Park plant. In 1952 Tasty began a five year, $8 million capital improvement program to mechanize and modernize its facilities. These efforts resulted in the reduction of production time from twelve hours to forty minutes. Even today—with four million cakes and pies produced daily—the plant is still not working up to its full capacity.

In 1968, when the neighborhood surrounding Tasty looked as if it was to become a slum, the plant and its facilities were worth $30 million dollars. Paul R. Kaiser, then Chairman of the Board, summed up the feeling of Tasty management to changing neighborhood conditions and the company's reaction to them in a speech to the American Newcomer Society:

> At that point [1968] the replacement value of the bakery's equipment and machinery was about $30 million, hardly the kind of investment you want to see go down the drain in a neighborhood of declining real estate values.
> We felt that if real estate plummeted in our part of Philadelphia, then Tasty Baking real estate would go right down with the neighborhood. Tasty management felt that looking out for its real estate was as much a part of its responsibility as protecting its stockholder's investment and making profits.

Barbara Baker: Campaign for Human Development, U.S. Catholic Conference—used with permission

Demonstrators show their concern for neighborhood preservation in Northeast Philadelphia, May 18, 1981.

Basically, then, it would have cost Tasty Baking Company a lot of money to move, had they decided to do so. Further, by moving Tasty might have had to bring in outside people to help finance relocating and plant building. Bringing in outsiders might have resulted in the relinquishing of some control by the current owners.

In addition to the economic costs of moving, Tasty also considered the impact of moving on its workers. The Hunting Park plant employs approximately 1,800 workers. Many of these workers had worked for Tasty for a long time; in fact, a large percentage of the work force is represented by several generations of families, many of whom had originally lived in the neighborhood surrounding the plant. Prompted by changing neighborhood conditions, the families of Tasty employees had moved to other communities nearby or to the northeast sections of the city. The workers were becoming somewhat fearful of coming to work, because of increases in crime and delinquency in the area. As one company spokesman said, "Happy workers make good cakes." Accordingly, Tasty was concerned about maintaining the high quality of its work force.

Tasty Baking Company has enjoyed a very stable work force owing to the excellent benefit package and salaries paid to employees. According to a company spokesman, Tasty workers are paid salaries which are "equal to or higher than the prevailing industry rate"; workers also receive two raises a year.[1]

The workers, however, are not represented by a union. The first attempt to unionize Tasty employees occurred in the late 1930's when the Teamsters tried to organize Tasty salesmen. In 1968 the American Bakers and Confectioners' Union also attempted to organize a union within the company. In the latter instance, Tasty petitioned the Labor Relations Board for an election, and the workers voted against unionization. A move to a different location might have resulted in a new work force that might not have enjoyed as good a relationship with management as the current one does. Tasty Baking Company, unlike other companies who faced a relocation decision, was not trying to flee from its workers or a union. These factors, and the cost of relocating, resulted in Tasty Baking Co.'s decision to remain in Allegheny West.

Tasty's decision to stay in Allegheny West was predicated on the belief that it must undertake some constructive action to stop and reverse the trend of neighborhood decline.[2] This action was in the form of a program of private capital investment in the community. The primary reason for relying upon private initiative was that in 1968 there was no public money available. The city government had no economic or housing development plans for the neighborhood. However one ultimate goal of the reinvestment was to establish eventually a triangular effort involving neighborhood residents, other local businesses and the city. Accordingly, the development of the Allegheny West Community Development Project (AWCDP) proceeded in several stages.

As a first step, Tasty commissioned the Jackson-Cross Company to survey the "Allegheny West" area.[3] The boundaries of Allegheny West are 22nd street to the east, Ridge Avenue to the west, Clearfield street to the south, and Westmoreland street to the north. The community, with a population of 23,300, was almost entirely residential, with approximately 6,300 single-family, two-story, row houses. The latter fact was important for the success of the program since the fostering of home ownership and property maintenance would be facilitated by just such a housing structure.

In addition to reporting on the type and quality of neighborhood housing, the Jackson-Cross survey outlined several neighborhood problems and their causes (vandalism, trash in streets and alleys, low civic interest, noise, insufficient property maintenance), causes of neighborhood change, and a general list of recommendations for dealing with the existing problems.

Tasty contributed $40,000 for the first year's trial effort. Almost all the money was used for administrative costs. However, Tasty did not want to run the program itself because they thought that this would be viewed as a paternalistic action of a big business. Instead, they sought someone from outside the company. Philip Price, lawyer, former public defender, and current Pa. State Senator, was recommended by Chamber of Commerce President Thatcher Longstreth and hired as the project director. Two years later, in 1970, Walter Evans, an engineer-builder, was hired to solve some of the technical problems associated with house vacancies. The project itself was administered by the Greater Philadelphia Foundation, the charitable unit of the Greater Philadelphia Chamber of Commerce. For the first two years Phil Price did block and community organizing, getting to know the neighborhood, the residents, and the various community groups.

Housing became the first priority activity of the AWCDP. In 1970 the AWCPD began to acquire and rehabilitate vacant properties. This undertaking was initially made possible by a mortgage agreement with the Philadelphia National Bank (PNB). Each house acquired was completely redesigned with all major structural faults repaired. New appliances were donated by Sears, Roebuck and Company, one sponsor of the project. To provide employment opportunities for residents, all repairs to the property were made by local labor. The project assumed responsibilities for structural defects for three years after sale.

The AWCDP took a loss on each vacant house that was rehabilitated and sold. The average price of a vacant house was $600, with rehabilitation costing approximately $14,000. The average selling price was $10,000—representing an average loss of $4,000. Mortgages were for twenty-five years under an installment sale agreement. The project was to maintain title to the house for the first three years after sale and was, therefore, subject to the risk of loss should the buyer default on his mortgage payment. The default rate to date has been minimal. Currently, financing is made available by four banks: PNB, Fidelity Bank, Girard Bank, and Central Penn National Bank. The availability of housing is publicized by word of mouth and public notices throughout the community. Community leaders then refer persons to the program.

Soon after property acquisition and rehabilitation activities began, the AWCDP initiated the second phase of its housing program: "Operation Facelift." This program provided technical assistance to existing homeowners to help them maintain their properties. In 1973 mechanics were also made available to residents to perform home-remodeling work at much reduced rates. Since 1970, approximately 160 houses have been rehabilitated and sold or rented.

In addition to the housing rehabilitation and maintenance services of AWCDP, the project initiated and funded a number of social service-oriented programs. These included a Montessori school and day-care centers for working parents; a "block" Scouting program; development of vest pocket parks and recreational areas; employment programs (e.g., summer job placement for area teenagers); and free legal advice and counselling to community groups and individuals.

The participation of community residents, local businesses and the city government in AWCDP became greater as initial project activities took shape. At the present time Allegheny West has several civic associations, a small businessmen's association, and approximately sixty block organizations. Twenty percent of the residents participate in the project in some form. The city became involved by providing better municipal services, including street cleaning, trash and garbage pick-up, more police patrols, and $48,000 for improved street lighting. Federal and Community Development money became available to the area in 1976.

One of the unique and important aspects of the project is the financial support it has received from business and manufacturing firms in the area. Although Tasty Baking Company still remains the largest contributor to the project, having contributed approximately $1.5 million since the project began in 1968, a number of other companies have also made financial and product contributions towards the goal of neighborhood revitalization. These include: Morris, Wheeler & Co., Penn Fishing Tackle Mfg. Co., Sears, Roebuck & Co., Philadelphia Electric Co., The Budd Co., Fidelity Bank, Girard Bank, Steel Heddle Manufacturing Co., Rosenau Brothers, Inc., Philadelphia National Bank, and Container Corporation of America.[4] Contributions from these companies amounted to approximately a quarter million dollars over the last decade.

The companies that have contributed to AWCDP have done so primarily at the urging of Paul Kaiser. Mr. Kaiser has continuously met with business leaders and sent letters to neighborhood companies explaining the nature of AWCDP and the importance of the Project in fostering community stability and economic vitality. Both Tasty Baking Company and others who contribute to AWCDP receive federal and

State tax benefits for their financial support. For example, since 1968 the state of Pennsylvania, through the Pennsylvania Neighborhood Assistance Act, has extended tax credits to companies that spend money to improve declining neighborhoods. Up to 1976 these credits amounted to 50 percent of the money donated to AWCDP; after 1976, the credit amount rose to 70 percent. The combined federal and state tax credit benefits result in a company being able to deduct approximately .90¢ of every dollar contributed to AWCDP. For Tasty Baking Company—after tax benefits are applied—the money contributed to Allegheny West accounts for less than one penny on each share of the company's stock. Thus, involvement in AWCDP has both positive social benefits for the community and its residents and economic benefits for the businesses that support the Project's programs.

In 1974 the AWCDP became the Allegheny West Foundation (AWF), a public charitable, non-profit foundation. The project, while still focused on housing and social service activities, is more closely overseen by neighborhood residents. The triangular structure of effort is still in effect; the board of trustees of AWF include representatives from neighborhood civic associations, city government, and area business people. The accomplishments of the AWCDP/AWF have been impressive, given their modest beginnings. The economic benefits to the community are in the amount of approximately $3 million dollars a year in direct funds and/or services.[5]

The decision by Tasty Baking Company to stay in Philadelphia and the efforts undertaken by the company in Allegheny West stand in sharp contrast to the actions of the business and manufacturing concerns that have left Philadelphia and other cities like it. In what ways was and is Tasty different in comparison to these companies? These factors should be noted:

Tasty has had a local reputation and orientation. Although its products are distributed in several states, Philadelphia is still its primary market. Its major plant operations are located in Philadelphia. The company was extremely satisfied with its physical plant facilities, having invested considerable capital in them. Moving from its Hunting Park location would have cost a great deal of money and a loss of control by the current owners. Tasty was also happy with its work force. Tasty was not fleeing the high salaries or demands of its employees.

The efforts of Tasty Baking Company show that private investment in one's community can work and that the costs need not be great. Certainly, if a "small" company like Tasty can impact on a situation of neighborhood decline, other firms can as well.

Notes

1. The company refused to disclose information regarding the average salaries of its employees.

2. Tastykake, more than most companies, had a reason to be sensitive to community opinion. In the early sixties, the company had been boycotted by a consortium of black ministers, led by Rev. Leon A. Sullivan, in Philadelphia because of its racially discriminatory hiring practices. The boycott effectively cut down sales of Tastykake products to Philadelphia blacks, major consumers of their wares. Tastykake hired blacks and has been responsive to community issues ever since. The boycott also pushed its originator Leon Sullivan into public prominence, and the prestige he gained helped him launch the international OIC training programs.

3. It was the Jackson-Cross survey that coined the name.

4. It is interesting to note that Container Corporation, now owned by the multinational Mobil Corporation, which has no local ties to Philadelphia, has decided to close down one of its plants in Manayunk (see section "Citizens and Unions Respond").

5. ALLEGHENY WEST FACT SHEET, 1977:

- 120 houses renovated and sold
- 6 mini-parks/gardens constructed
- 6 church programs assisted
- 60 small businesses aided
- 1,000 summer jobs for kids
- 1,700 boys involved in Boy Scouts (including summer camp)
- 15 houses renovated and rented
- 2 day care centers established
- 3 civic association headquarters established
- 1 branch library renovated
- 150 residents employed by local business
- Vastly increased city services (systematic street and sidewalk repair, abandoned car removal, new and improved street lighting, etc.)

ECONOMIC IMPACT ON THE COMMUNITY

135 renovated houses:
- Relieves HUD of $15–20 daily costs (inspection, record-keeping, etc.)
- Returns $40–50,000 to the city annually (real estate, sewer, water taxes)
- Ranges the income per house between $15,000 and $25,000, since in most cases husband and wife both work.

This pumps at least $1,800,000 into the community each year, and this impact, in turn, attracts about $1,000,000 more in services–gas stations, barber shops, cleaners, and small retail stores to serve a growing population:

- 3 fast food outlets: Gino's, 7-11, Church's Fried Chicken.
- A car wash and sports store owned by Eagles' Harold Carmichael, which provides jobs for neighborhood kids.
- A 96-unit, $2.5 million cooperative townhouse at 21st and Somerset Streets.
- Total restoration of the 22nd Street business district. Nearly dead ten years ago, every store is operating today. Off-street parking is now provided for shoppers.

Because of government grants, which pay almost all administration costs, 95 cents of every after-tax dollar invested by private businesses goes directly into projects and programs in the Allegheny West community.

Chapter 11

Citizens and Unions Respond

Pamela Haines and Gary Klein

Philadelphians have not gone gently into the economic good night. Their unions have negotiated against plant closings, filed complaints with the National Labor Relations Board, lobbied for legislation to aid workers affected by plant relocations and organized and participated in protest demonstrations and boycotts. Most significantly, although the city's unions played key roles in all of these actions, they did not act alone. They formed effective coalitions with church groups, activist citizens and sympathetic political leaders.

The city's unions have a long history of social and political activism. The Amalgamated Clothing Workers (ACWA) has pioneered in health, housing, and old age benefits for its members since its inception in 1929. After World War II, it became active in politics, conducting the first legislation drives among blacks and Puerto Ricans and encouraging members to run for Democratic Party posts. The Amalgamated was closely associated with the city's Reform Movement and reform mayors Joseph Clark and Richardson Dilworth (1951–1962). In addition, the Amalgamated supported the struggles of other unions such as the United Farm Workers.

In the early 1970's, with declining membership and increasing need for assistance itself, the Amalgamated sought to broaden its base of support. Amalgamated leaders realized that the city could never hope to revitalize its clothing industry and provide work for union members if textile and clothing manufacturers could hire non-union workers at significantly lower wages in the South and Southwest. Thus the Philadelphia Amalgamated eagerly joined with other locals across the country to aid the unionizing efforts in the South and Southwest.

In 1972 Farah was the largest manufacturer of men's pants in the world. Its Chicano apparel workers in Texas went on strike after the company refused to negotiate with the elected union. In attempts to break the newly formed ACWA locals, Farah had tried to discharge union leaders and in other ways intimidate union loyalists. A national boycott was organized. It succeeded in tempering Farah's ways. The Philadelphia Amalgamated actively participated in the boycott, and was an important part of the union's success in Texas. In return for help in the boycott, the local Amalgamated supported its church, community, and political allies in city-wide coalitions on such issues as the Vietnam war, jobs and energy.

On the Friday before the 1976 Presidental election, ACWA members attending a Carter Rally "put down their banners and began to distribute 'Boycott J. P. Stevens' leaflets at thirty center city locations. It was the opening move by the Philadelphia clothing and textile workers in support of a national effort to force the Stevens Company to recognize the union in their southern plants."[1]

On November 30th of that year, a meeting of clergy, labor, and community leaders was held to explain the reasons for the boycott and call for their support. Within a month, the endorsements included eighteen from clergy, thirty-three from community and civic groups, forty-two from labor, and seventeen from state representatives and congress people.

The list continued to grow. In early December the union bought shares in Stevens stock, and in March twelve staff members attended the annual stockholders meeting in New York City. Joining 2,500 others, they demonstrated for company accountability on labor relations and minority hiring practices.

Up to this point the union had provided virtually all the initiative and energy for boycott activities. By the spring of 1977, however, the "Citizens' Committee in Support of Justice for J. P. Stevens Workers" had been formed. It took over responsibility for Philadelphia boycott activities. Formally independent of the union, it included clergy, labor, students, civil rights workers, community and political leaders. Members of the Philadelphia congressional delegation, the mayor, and representatives of the Eagles football team helped lend prestige. In such a strong union town, participating in the boycott was seen as patriotic. While involvement of union membership on the whole was not high, the union continued to play a vital role in the Citizens' Committee through its boycott staffperson, Hy Goldberg.[2]

The central activity in Philadelphia was the attempt to get store owners to cancel or cut back their orders of Stevens linen products. Initial meetings between members of the Citizen's Committee and store owners were to set up the moral issue, to demonstrate that there was no possible neutral stand. If compliance was not forthcoming, the meetings were followed by leafletting to encourage a consumer boycott of Stevens products in the store, then more negotiations with the owners.

Among the big chains, the first target was Wanamaker's, which, in the face of such widespread support for the boycott, agreed quickly to cut its orders way back. Attention then turned in mid-1977 to Gimbels, where initial meetings were less fruitful. Regular leafletting outside the store continued into 1978, along with creative publicizing of the boycott within. Members of the Citizens' Committee, wearing T-shirts emblazoned with the bright red message: "Boycott J. P. Stevens—sheets, towels, carpets, table linens, fabrics," would take an hour or two to shop at Gimbels, spending much of their time browsing through the linens department. Although the campaign had significant impact on linens sales, the national Gimbels chain was intransigent on cutting back their stock, and negotiations dragged on till the boycott's end. Actions were also taken against Strawbridge and Clothier, Pomeroy's and smaller neighborhood stores throughout the city. By the fall of 1980, J. P. Stevens products had been cut back by about 85 percent at Wanamaker's, 75 percent at Gimbels, 95 percent at Strawbridge and Clothier, and 85 percent at Pomeroy's.

In addition to publicity through department store leafletting, the committee arranged public viewings of the boycott film, "Testimony." The film drew 1,000 people to a center city rally and march in support of the boycott in the fall of 1977. A special screening of the very timely, popular movie, *Norma Rae*, was sold out in mid-1979. This showing featured an introduction in support of the struggle by union leaders and a demonstration outside.[3]

The third prong of the strategy, the corporate campaign, found its form largely in local support demonstrations for actions in New York City and in some trips to join those actions. In New York boycotters attended board meetings of corporations having "interlocks" with J. P. Stevens. During the meetings demonstrations were held outside in support of resolutions being offered inside. The goal was to raise hell and generate adverse publicity, in the hopes that other corporations would be embarrassed by their ties with Stevens and be glad to be rid of this troublesome bedfellow. Over the course of the campaign, New York Life Manufacturers Trust and Avon Products did terminate their boardroom ties with Stevens.

In October of 1980, attention was focused on the interlocking directorates of Sperry, J. C. Penney, and Metropolitan Life. The latter, provider of 65 percent of J. P. Stevens working capital, was a particularly crucial link in the chain of corporate support. Less than a week after the October demonstrations began, Stevens announced its willingness to yield to the union's right to represent the workers at the Roanoke Rapids textile plant.[4]

While it is hard to evaluate one piece of an overall campaign in isolation, it is clear that Philadelphia boycott activities had significant impact on the local stores. They also mobilized a wide cross-section of people—a vital church/community/academic/political/labor coalition—and educated even more. Hy Goldberg, boycott coordinator from union headquarters in New York City, in a fall, 1980 interview, commented on the high degree of enthusiasm and involvement in the Philadelphia campaign.

Nationally, of the three prongs of the strategy, the consumer boycott probably had the least direct impact. The Stevens linen department lost about 12 percent of its sales as a result of the boycott. While this decline amounted to millions of dollars, a 12 percent loss in a department consitituting only 35 percent of the entire business was relatively insignificant in terms of overall profit. Stevens could have absorbed such a loss for years without being forced to its knees.[5]

The combination of the national boycott with the work from the South in exposing Stevens' labor record, however, was a powerful means of educating people on the legitimacy and importance of labor struggles and undermining external

Pennsylvania Public Interest Coalition

Pennsylvania Public Interest Coalition

Lobby Day in Harrisburg, Pennsylvania, October 6, 1981, organized by the Pennsylvania Public Interest Coalition. The theme was "Save Pennsylvania's Economy." *Top*: James Farmer acknowledges the response to his speech.

support for the company. By using the boycott to focus on the moral issue of the right to organize, the campaign kept Stevens in the glare of unfavorable publicity.

Even this concentrated public disapprobation might have been endured for a long time by such an avowedly anti-union company as Stevens. When the focus widened to include Stevens' corporate interlocks, however, the company's real vulnerability was exposed. It could not exist in corporate isolation. When public pressure became so great that the best self-interest of Manufacturers Hanover Trust, for example, lay in dumping the Stevens chairman from its board, the campaign began to have decisive impact.

The boycott was won. After four and a half years of steady work, and $3.5 million a year from the Amalgamated, the wedge in solidly anti-union southern textiles had been made. The Amalgamated has begun talking of stepped-up organizing at other Stevens plants, and other unions, such as the IUE, the UAW and the Furniture Workers, are shifting more of their attention to organizing in the South. But it is a limited victory.

The outlook for Philadelphia clothing and textile workers has not changed. There is little chance that the industry will move back north again unless southern wages go up significantly and/or transportation energy cost increases shift the economic advantage back to the northern ports and points of distribution. In this respect, the J. P. Stevens campaign in Philadelphia was an example of closing the gate after the horse was gone.

In the larger context, the campaign will have on-going impact only to the extent that it is followed by equally vigorous and ever-widening struggles to equalize the conditions and strength of labor all over the country (and, by logical implication, all over the world). While such a response by labor and its allies to runaway shops provides no easy solutions, the example of the Stevens boycott did provide some lessons and directions for the future:

1 The ability of northern unionists and southern workers to see each other as allies in a larger struggle rather than contestants for the same small piece of pie, and the willingness of many Philadelphians to work for something that wouldn't yield immediate local benefits greatly strengthens their hand in challenging the mobility of profit-seeking corporations.
2 A single, national focus helps to concentrate energy. At the same time, a carefully developed multi-pronged strategy allows many people to participate in different ways and avoids the danger of having all the eggs in one basket.
3 The potential for building broad-based coalitions is greatly increased when labor is able to define its struggle in broader terms than contract negotiation.

That these lessons were taken to heart was demonstrated by the formation of the Delaware Valley Coalition for Jobs on Dec. 18, 1979. The Coalition, whose members run the ideological gamut from radical to conservative with a large liberal core, numbers among its members over fifty union, church, and community organizations from Philadelphia. They have worked together to organize pickets, marches, public hearings, press conferences, and lobbying efforts to oppose plant shutdowns and the disregard of big business for the social and economic well-being of working people.

The Coalition was officially formed when representatives of the United Auto Workers (UAW) and the Philadelphia Unemployment Project (PUP) arranged a meeting attended by forty leaders of local union, church, and community organizations. The organizing meeting was called because of the increased amount of activity in fighting plant shutdowns and the realization by UAW and PUP leadership that much more such work would be needed in the near future. The staff of the Coalition derived in large part from PUP, and office space and supplies were shared in order to hold down costs and reap advantages from an established organization with similar, but not identical, purposes.

When the Coalition originated, some of its member organizations were actively engaged in a battle with ITE Gould. Gould, a major conglomerate, bought ITE Imperial, a 100 year old Philadelphia–based company in the mid–1970's. Soon after the purchase, Gould gradually and secretly moved equipment from their circuit–breaker plant at 19th and Callowhill Sts. to a new plant in North Carolina. Gould kept this secret in order to keep up worker morale and productivity. It also sought to avoid receiving sanctions for breaking an arbitrator's decision, as had happened in June 1969. The idea was to keep work at the plant until after the expiration of the company's contract with UAW Local 1613 signed in 1977, which ended on March 28, 1980.

Joseph Ferrara, Assistant Director, region 9 of the UAW, and president of the DVCJ, testifying before the U.S. Senate Committee on Labor and Human Resources on Oct. 29, 1979, detailed his experience with ITE Gould:

> Soon after Gould acquired the company, a new collective bargaining agreement was negotiated in April, 1977. It was a very good agreement, providing equity for our members. It was in most respects on a par with our pattern "Big Three" auto contracts—with one exception. There was no supplemental unemployment benefits plan because there had never been any layoffs.
>
> The ink was barely dry on that agreement when there began to be persistent rumors that Gould was building new plants in the South, to which it was planning to relocate large numbers of our members' jobs. The corporation–in the person of ITE's former

board chairman, now a Gould vice president–flatly denied the rumors.

Acting on a tip from office employees, the president of the local union and I went to Wilmington, North Carolina in September 1977, where we discovered a Gould plant in preparation. Management employees at the site confirmed that Gould was intending to relocate small circuit–breaker production from Philadelphia to that new plant. Subsequently, we discovered through the same kind of detective work that other new plants were in various stages of preparation in Florence, Casey, and Woodruff, South Carolina; and that switch gear plants—the mainstay of Gould ITE's Philadelphia-area production—were being prepared in Sanford, Florida and Tulsa, Oklahoma. As of this date, employment in the Philadelphia area has already dwindled to 2,000, a decline of one-third since Gould acquired the business. For now, those remaining jobs are being preserved thanks to an arbitrator's ruling that Gould's removal of jobs and equipment constituted a violation of the provision I mentioned earlier of our collective bargaining agreement. Pursuant to that arbitrator's decision, in June 1979 Gould had to return all of the equipment it had removed from its Philadelphia-area plants. But our collective bargaining agreement expires next spring, and I don't mind telling you we are worried about the future. If we had lost that arbitration case, the company would have closed, and the remaining 2,000 jobs would have been lost.

The attitude of Gould's top management toward unions is not reassuring. That attitude is typified by the following statement of Gould's board chairman and chief executive officer, William T. Ylvisaker, to the New York Society of Security Analysts in 1978: "We don't like unions. . . . I don't mind saying that. I tell them that. We fight them. We've had two decertifications of them last year, as a matter of fact. Two unions had to leave; the employees voted them out. And we've only had one plant, previously not unionized, that has been organized during my tenure at Gould."

Such flagrant disregard for workers rights underscores the need for federal action to provide meaningful job security protection and to mandate corporate responsibility for assistance to workers in the event of job loss. In the face of unbridled corporate power to move plants to avoid unions and increase profits, collective bargaining alone often cannot do that job.

Throughout long years of bargaining with the ITE, that corporation never claimed that its Philadelphia operations were losing money. Only in the past year has Gould claimed that one of three local former ITE divisions—Small Air, which makes small circuit–breakers for home use—was losing some $400,000 annually *after* a $15 million "corporate charge" which I take to mean under the old formula . . . ITE would have shown better than a $14 million profit.[6]

Gould is not a newcomer to plant closings in the Philadelphia area. Just before the Pension Reform Act (ERISA) took effect, Gould closed its former piston ring plant in southwest Philadelphia (UWA Local 585). At one time, that plant provided jobs for 600 workers; employment had dwindled to about 200 at the time of the shutdown. We had to go to court to force Gould to arbitrate a dispute over its debt to our members' pension fund.

Dealing with Gould has been a frustrating experience for me as a trade unionist. That corporation's evident disregard for the welfare of its Philadelphia workers, who gave long years of faithful service to ITE, is callous, to say the least. I am convinced that Gould's apparent plan to move production and jobs out of the Philadelphia area is based on greed, not economic necessity. Gould is not a Chrysler, in dire financial straits—far from it. It is a highly profitable, consistently profitable and fast growing corporation. Apart from the figures mentioned earlier regarding the Small Air Division, I do not know what Gould's (or ITE's) profits from local operations have been. That is not a subject which either company ever elected to discuss with me. But I do believe there have been profits. Perhaps not enough profits to satisfy the inordinately high return on investment target of head–office planners; perhaps not as high as would be made, for a few years, at new southern plants—plants which may have been built in part with taxpayers' money in the form of industrial revenue bond financing. The federal government would also do its part to underwrite any move: a favorable tax code will allow write–off of the old plants, accelerated depreciation of the new plants, investment tax credit towards the purchase of new equipment, and deduction of any relocation costs as ordinary and necessary business expenses. To make matters worse, federal law sanctions interstate differences in unemployment insurance and workers' compensation rates, and permits state laws to bar union shops.

I am convinced that the southern workers will be only temporary beneficiaries of our misfortune. For a company like Gould, North Carolina is only a temporary stopover on the flight to Singapore, Mexico or Taiwan; management officials of that company have told me as much. If and when that happens, the federal tax subsidy will be even greater than if the relocation were inside the U.S.—as a result of the foreign tax credit and deferral of taxes on unrepatriated profits."

On December 3, 1979, three months after Ferrara's testimony, approximately 150 retired UAW members plus other union and community people organized by DVCJ picketed the Gould plant to protest the removal of equipment to North Carolina, the failure of the company to bargain in good faith with the employee's union, and the plan to move 2,200 jobs south. None of the Gould employees or other UAW members picketed due to legal constraints. The event was well covered by the press and added pressure on Gould to change their original plans and maintain most of the workers' jobs. Gould was acting out of self-interest; it was afraid of the ramifications of further legal and public pressure. Even so, it was rewarding for the picketers to see that a company's self-interest could be changed to maintain instead of delete local manufacturing jobs. The Gould "victory" became the major rallying point for the DVCJ. It was proof that coalition can make a difference. For this reason the Gould site was chosen as the destination for a Coalition–sponsored march from City Hall on March 15, 1980 to dramatize the problem of job loss and to increase public awareness of, and support for, state and federal plant closing legislation. The march attracted over 350 participants.

Between the picket at Gould and the march from City Hall to Gould, the Coalition, in collaboration with PUP, sponsored a public hearing at City Hall, titled "To Save Our Jobs and Our Neighborhoods." It occurred on February 16, 1980, and attracted an audience of approximately six hundred, which were for the most part union members. The twenty speakers included union leaders, community group leaders, and local politicians, who spoke to elicit support for state and local legislation that would protect workers from runaway shops and plant closedowns.

Ten days after this event, DVCJ held a press conference in the State Capitol with other Philadelphia organizations actively supporting HB 1251. This bill requires companies to give at least one year's notice to unions and the community if a plant is planning to close, six month's continuation of health care benefits to the employees' families after closing, severance pay, and financial aid to the affected community.

The stated purpose of the rally was to promote a discharge petition for the bill so it could get back onto the House floor, but the real intention was to attract publicity for the bill since the petition had no chance of passing, needing a two-thirds vote to pass when it would be very difficult to obtain even a majority.

This activity was consistent with DVCJ's coordinator John Dodd's philosophy in appoaching HB 1251: "The struggle is as important as what we get out of it. . . . The point is to build an awareness in people that we need more protection from corporate power. . . . If we get this then we go for more."[7]

While working for the passage of local and state laws, Coalition leadership never lost sight of the need for national legislation. Testifying before the Senate Subcommittee on Labor and Human Resources, Joseph Ferrara of the UAW discussed the conditions that made early warning and worker compensation necessary on a national level.

> A number of other UAW organized plants in the Philadelphia area have been closed in recent years.
>
> When Strick Trailer closed its Langhorne, Pennsylvania plant in June 1975, some 1,500 UAW members lost their jobs. That major shutdown followed close on the heels of Zenith's announcement of plans to close its Lansdale, Pennsylvania home entertainment plant (acquired previously from Philco-Ford), throwing eight hundred people out of work. To add insult to injury, the Department of Labor rejected our petition for trade adjustment assistance, despite the fact that Japanese imports were a major factor in the shutdown decision. The Simonds Abrasive Division of Wallace Murray Corporation closed a plant in our region several years back, idling four hundred workers; the same parent company closed another plant in our region, Murray Body. Quaker City Iron, which made large gas tanks for gas stations and others, is yet another plant of two hundred which closed recently. It was owned by two brothers in their seventies, who wanted to retire and claimed they couldn't sell the plant; many of their former employees still have not found work.
>
> The Budd Company is another major Philadelphia-area employer where our members have serious concerns about job security. This company operated what was at one time the largest machine shop in the U.S.—and the largest stamping plant in the East. Employment has dwindled to the point where we now have more Budd retirees in our union in the Philadelphia area than active employees.
>
> Budd is the nation's only remaining producer of rail cars—an

industry which should be undergoing renaissance and revitalization in these energy-short times. Yet that favorable economic prospect has not filtered through to our members at Budd in the Philadelphia area.

The Budd Company's Philadelphia operations built the dies for that West German auto maker. When Volkswagen opened its new U.S. assembly plant in New Stanton, Pennsylvania, our members at Budd didn't even have the chance to bid for the die work for that plant; the contract went to a Japanese company instead.

Like ITE, Budd was recently acquired—in 1978, by Thyssen, a giant West German steel corporation. And, in a similar fashion, soon after the acquisition we learned that Budd had taken options on a huge former Celanese Corporation plant in Rome, Georgia, where it was planning to relocate Philadelphia-area jobs and production. I visited that Georgia site for the first time last week. Already last year, Budd closed its main press plant in Hunting Park, Philadelphia, at a cost of 1,200 jobs.

It is ironic, and shocking, that this corporation is playing fast and loose with thousands of jobs in a way that would never be tolerated in its parent company's home country. Could you imagine Thyssen closing plants and idling thousands of workers in one part of Germany, while relocating production to new plants in other parts of the country? It would be unthinkable, politically, socially, from every standpoint. In West Germany and other European countries, there is a wide range of laws and policies to protect workers and communities against economic dislocation.

What I have related is the experience of my union. Plant closings in the Delaware Valley have hit other unions even harder—the Steel Workers, Rubber Workers, Electrical Workers, and Clothing Workers, among others. The town of Conshohocken, Pennsylvania was literally built around Alan Wood Steel, which went out of business in 1977 with disastrous consequences. That town is also the home of Lee Tire—now part of

Goodyear, which is closing that plant and relocating production to a new facility in Oklahoma City.

There are few decisions in our modern industrial society with impact more profound than corporate decisions; corporations are presently under no legal obligation to scrutinize anything except the projected impact upon their 'bottom line.' Indeed by the tax code and in numerous other ways, the government presently subsidizes even the most callous and irresponsible of corporate shutdown and relocation decisions. The present situation in our country today can only be described as "industrial piracy." Playing on fears of job loss, corporations pit one community against another, one state and region against another, in endless pursuit of tax abatements, subsidies and other bribes, undermining job security and eroding the tax base everywhere without adding a single new job nationwide.

Meaningful public restraints on corporate behavior in shutdown and relocation situations and an end to senseless industrial piracy are long overdue. Comprehensive federal legislation to achieve these goals is urgently needed.

Corporations, after all, are chartered by society. That charter does not give them the right to maximum profit free from any public restraint. Government has a right—and an obligation—to enact standards of decent corporate behavior. Plant closing is an area where such standards are urgently needed, and long overdue.[8]

To reenforce public perception of the need to enforce moral and societal standards on business corporations, the Coalition continued to schedule public events. On September 19, 1980, approximately one hundred fifty demonstrators, including South Philadelphia clergymen, DVCJ members, off-duty Du Pont workers, Du Pont workers on their lunch breaks, members of their families, and employees of other firms picketed in front of Du Pont Company's paint plant in Southwest Philadelphia. The demonstration was incited by the plan to phase-out three hundred jobs at the plant and open a new plant in Virginia. Du Pont decided to move South in spite of an offer by the city of Philadelphia to provide free land for them in an industrial park with 100 percent tax-exempt financing on building and machinery, as well as other significant tax inducements. Although unable to keep the paint plant in Philadelphia, the Coalition saw that the demonstration and other related activities were important in raising public awareness of the larger issues. They held Irving Shapiro, chairman of the DuPont board, responsible for the plant closing decision. They sought to point out the contradiction between his decision to close this profitable plant and his position as head of a national labor-management panel created by President Carter to find ways to halt urban industrial decline.

On November 12, 1980, the general manager of the Folding Box division of Container Corporation (CCA) in Manayunk, announced plans to idle one hundred nineteen workers in the Folding Box plant and to issue the final decision on December 15. CCA has been owned and directly governed by Mobil, the fifth largest corporation in the world. Profits accruing to Mobil from CCA had been extremely high. In 1979, Mobil earned $46 million from CCA, a 228 percent increase from their 1978 earning of $14 million. In fact, Mobil received a profit from this very plant they planned to close down. However, they anticipated bigger profits from closing this plant down and modernizing others. DVCJ rapidly investigated the situation. They discovered the city had twice accommodated the manufacturers in 1979 when they sought special help for easing pollution standards and for a low interest loan for a new plant. There had been verbal guarantees that in return for this aid there would be no job loss. Realizing that Mobil, a corporate giant, could easily maintain this plant, even if it did become less profitable in the future, the Coalition decided to go all out to protect this plant and the workers' jobs.

On December 4, the Coalition, in collaboration with United Paperworkers International Union Local 392, scheduled a rally at the union hall in Manayunk. Afterward, organizing at the union hall in the middle of the afternoon, seventy-five people blocked a major traffic intersection and dispersed leaflets to rush hour motorists. The leaflet summarized the facts concerning the tentative shutdown and advertised a public meeting in neighboring Roxborough.

The town meeting was held on December 11, with an audience consisting of approximately two hundred employees, their families, union members, and friends and neighbors of the employees. They listened to several speakers with union, political and religious ties. Joe Pedrick, the President of Local 392, pledged to take the battle "to the labor courts all the way up the line until we are satisfied with the answers we get from CCA management and from the Mobil Oil Company".[9]

A clever ploy was initiated by Pedrick and the Coalition to generate further awareness in the Manayunk/Roxborough district and put moral pressures on the person who officially made the plant closing decision—Richard Graham. Twenty-three hundred Christmas cards were distributed to community residents through the local churches to send to Graham which stated "Happy Holidays from the workers and families of CCA and Manayunk-Roxborough. Don't close our plant—We need these jobs."

As a result of these two public displays of support for the workers, the Paperworkers' Union had more strength in negotiations with management, and a final decision was postponed from the originally stated December 15 until early

February. A separate Marayunk/Roxborough Coalition for Jobs was formed with the support of DVCJ. However, in spite of these attempts and a further attempt to influence the man at the top of CCA and Executive Vice President of Mobil Corporation, Richard Tucker, with a private meeting and personal letters from community residents sent to him, CCA announced its closing in early February of 1981.

There were good things that emerged from this battle, even though the workers lost their jobs. The workers were given an unprecedented amount of compensation; and the local community became aware of, and to some extent involved in, fighting exploitative corporate practices.

Before DVCJ employees had a chance to catch their breath from this struggle, they were plotting strategies to reverse the decision of Sun Ship in Chester to eliminate over 3,000 jobs. This work was being done with the realistic understanding that they would most likely only be able to further public awareness of what big businesses are doing to their manufacturing employees.

In the summer of 1981, the Coalition led a series of demonstrations in support of the workers at the Eaton Corporation's forklift plant in Northeast Philadelphia. Four hundred and fifty workers will lose their jobs at this plant as Eaton continues to shift production south and overseas. When severance terms were agreed to, the settlement proved to be one of the richest close-out contracts ever negotiated in the Philadelphia area. "There's no way we could have gotten these benefits without a major fight," said Danny Chmelko, International Association of Machinists' Business Representative.

Despite this settlement, the union and the Coalition intend to press their demand that E. M. DeWindt, Eaton board chairman, live up to his promise to "ease the impact of the closing" by moving a new Eaton product line into the city. A highly skilled workforce is available, and the city of Philadelphia has offered Eaton a five-year tax abatement, prime industrial land, and a publicly-financed plant.[10]

The efforts of the Delaware Valley Coalition for Jobs are now linked with other similar organizations across the state in the Pennsylvania Public Interest Coalition. On October 6, 1981, over one thousand community activists came together in Harrisburg, the state capital, for a Citizen Lobby Day to call for the passage of plant closing legislation and to support other measures in the interest of working people across the state.

James Farmer, long-time civil rights leader and now director of the Coalition of American Public Employees, expressed their hope and enthusiasm as he addressed the crowd there: "What you are doing here in Pennsylvania is being duplicated in state after state. We haven't had a coalition like this since the early civil rights movement of the '60's."

Notes

1. *Amalgamated News* (Nov.–Dec. 1976), p. 6.
2. *Amalgamated News* (Nov.–Dec. 1976); *Amalgamated News* (First Quarter 1977), p. 6.
3. Hy Goldberg, of the ACLU, interview.
4. *Amalgamated News* Fall 1977; Summer 1979.
5. Ibid.
6. Gould Inc. was listed 152 on the Fortune 500 list in 1978. Gould acquired ITE in 1976. At that time ITE had assests of $384,000,000 and a net worth of $199,000,000.
7. Andy Ziffer, "Corraling Runaways: As Jobs Go South Workers Organize," *Community Newspaper*, Feb. 21, 1980.

8. Campaigning for office at a Teamster's luncheon on October 10, 1979, Mayor Green promised to initiate such a program.

9. "Still No Firm Decision on Container Closing," *Manayunk Review*, Jan. 8, 1980.

10. "Eaton Corp. Is Forced To Pay Up," *Public Interest News* (Jan. 1982), p. 5.

Part Three

Community and Capital: The Ethical Issues

Chapter 12

Economic Development As If Neighborhoods Mattered

Edward Schwartz

If the main conflicts between cities and corporations in the 1970's revolved around issues of urban disinvestment—plant closing being the most dramatic example—the parallel conflicts of the 1980's most certainly will revolve around emerging strategies for urban economic development. At this point, the private sector has established a commanding position from which to set its own agenda in this area. The ease with which companies have relocated from central cities to suburbs, even to other parts of the country and the world, in response to unions and local taxes has brought mayors and city councils to their knees, literally begging corporate executives to return on any terms that suit them.

It is clear, moreover, that even stable neighborhoods are expendable in the process. The case involving construction of a General Motors Plant in a section of Detroit that encompassed a predominantly Polish neighborhood—Poletown, as it was known—is instructive. Detroit's Mayor, Coleman Young, hardly a reactionary, grew so tired of watching automobile companies move large plants out of his city that he volunteered to produce almost any package that would persuade them to remain. When in 1980 General Motors informed him that they might consider constructing a new Cadillac plant if the price were right, the Mayor got his wish. To accommodate G.M., the city cleared away 395 acres of land for the 70-acre plant, produced $182 million of public investment to "leverage" the $500 million of private investment promised by the company and offered a 50 percent real estate tax abatement for the first twelve years of the plant's operation. It also seized by eminent domain the entire Poletown neighborhood, with a population of 3,400, and demolished 1,000 homes, 16 churches, 3 schools, and 150 businesses—all to produce the 3,000 to 6,000 jobs that the new Cadillac plant would bring.

Mayor Young was not even apologetic about this undertaking. Indeed, he secured support for it from eleven of the twelve members of the Detroit City Council, Governor Milliken of Michigan, the Chamber of Commerce, the United Auto Workers, and the Catholic Archdiocese. "There's nothing wrong with our city," the Mayor announced at the signing of the G.M. agreement, "that two or three plants like this couldn't solve." Later, he defended his position to a CBS Reports interviewer in terms that mayors throughout the country would find familiar:

> The Governor of Ohio makes a foray into Michigan every year and tries to rip off some of our major industries and brags about it. If you're going to remain the competition, if you're going to play by the game, you've got to play by the rules as they are. And the rules as they are say that you offer concessions.[1]

To the Poletown neighborhood, however, the "reality" of the concessions was the destruction of their community. In an echo of a different period of our recent history, a part of the city had to be destroyed in order to save it.

Given current trends in economic development, neighborhoods throughout the country are wondering whether similar forms of salvation will be visited upon them. There is plenty of precedent, of course. The urban renewal programs of the 1950's demolished neighborhoods with great pride, "clearing" them for people closer to the planners' conception of decent living than the average impoverished urban villager was. In the battles between highways and ethnics in the 1960's, the highways won establishment support nearly every time. If high-rises and turnpikes were valid grounds to destroy neighborhoods in the 1950's and 1960's, then surely new jobs will provide ample justification in the 1980's. Mayor Young himself has laid the groundwork:

> If, let's say, that a city decides that a neighborhood is dispensable, then the rights of private property no longer apply. That's the whole concept of eminent domain. That's been embedded in our state and national constitution since the beginning of democracy, the theory of majority rule. Now if we're gonna' go by some new theory—if you're gonna' tell me that individual rights take priority over the rights of the whole in any situation, then you're dealing with anarchy. There's nothing revolutionary about what we've done. What is new is that the City of Detroit decides that jobs are in the public interest. Now the United States has decided time and time again that jobs are in the public interest, by the jobs program, CETA, and by others that the government has sponsored—WPA in the Depression. This is public purpose. There's no difference.[2]

One can hear mayors all over America making similar arguments, as the 1980's unfold.

In an important sense, Mayor Young was *not* espousing anything "new" or "revolutionary," especially within the framework of modern liberalism. To today's economists, even Keynesians, discussing neighborhood communities in connection with the larger economy is an anomaly, like talking about the Crganized Block and Air Traffic Control. Liberal models of politics pay little attention to neighborhoods as well. In the pluralist universe, there is the individual and the state, each in permanent tension with the other. The only collective force binding the two is interest groups whereby people who share similar characteristics and claims come together to make themselves heard within the political process. A neighborhood group can become an interest group, of course, but it deserves no special place in the panoply of groups with which liberal government must contend. In the end, politicians must balance all of these competing forces to achieve a "public good," conceived in terms of what seems best for each individual within the system. Within such a framework, then, destroying a neighborhood to create jobs is perfectly reasonable, as long as individuals appear to gain something in return.

In the broader spectrum of American politics, however, this particular balancing act has never been quite so clear-cut. Especially in earlier periods of our history, even liberal theorists and leaders recognized the distinct importance of local communities to a decent society. When the goals of economic development and neighborhood preservation came into conflict, they viewed the situation as a genuine, almost tragic urban dilemma, with pro's and con's on each side, instead of an open-and-shut case in favor of business development every time. The healthiest process, moreover, was viewed to be one in which economic development was planned with neighborhood preservation in mind. Acting on the basis of this understanding, city planners favored bringing neighborhood-based organizations directly into the economic planning process of the city. In working for social stability and political democracy, they believed that neighborhood preservation was a priority.

This position is rooted in our earliest traditions. Among the greatest advocate of local community organizations in America, in fact, was the founder of American liberalism, Thomas Jefferson. While many now believe that the principles of "life, liberty, and the pursuit of happiness" in the Declaration of Independence justify private autonomy and mobility, Jefferson himself had a quite different notion of what they meant. Happiness was not a function of self-aggrandizement, but of helping others, the instinct for which was part of an innate "moral sense":

> I believe . . . that it is instinct, and innate, that the moral sense is as much a part of our constitution as that of feeling, seeing, or hearing; as a wise creator must have seen to be necessary in an animal destined to live in society: that every human mind feels pleasure in doing good to another. . . . The essence of virtue is in doing good to others.[3]

Consequently, in politics, Jefferson believed that our national governmental system had to be built upon a foundation of local "ward republics," which would reflect the natural, communal alignments of the people:

> The article, however, nearest my heart is the division of counties into wards. These will be pure and elementary republics, the aim of all of which taken together composes the State, and will make of the whole a true democracy as to the business of the wards, which is that of nearest and daily concern. The affairs of the larger sections: of counties, of States, and of the Union, not admitting personal transaction by the people, will be delegated to agents elected by themselves, and representation will thus be substituted where personal action becomes impracticable. Yet, even over these representative organs, should they become cor-

> rupt and perverted, the division into wards, constituting the people in their wards a regularly organized power, enables them by that organization to crush, regularly and peaceably, the usurpations of their unfaithful agents, and rescues them from the dreadful necessity of doing it insurrectionally. In this way we shall be as republican as a large society can be and secure the continuance of purity in our government by the salutary, peaceable, and regular control of the people.[4]

To Jefferson, then, it was not isolated individuals competing with one another for wealth who would fulfill the republic's highest destinies, but high-minded citizens working together to achieve cooperative goals in local communities throughout the country.

The most penetrating foreign observer of early America, Alexis de Tocqueville, also recognized the central role that local communities played in shaping personal character and social values. America's success, he argued, grew not so much of out of our Constitution and laws, but our "customs," most of which were transmitted locally. It was customs that rendered us "the only one of the American nations that is able to support a democratic government." Indeed, it was, "the influence of customs that produces the different degrees of order and prosperity which may be distinguished in the several Anglo-American democracies."[5] Among the most important of these was the "Spirit of Township" that de Tocqueville found in New England:

> The New Englander is attached to his township not so much because he was born in it, but because it is a free and strong community, of which he is a member, and which deserves the care spent in managing it. In Europe, the absence of local public spirit is a frequent subject of regret to those who are in power; everyone agrees that there is no surer guarantee of order and tranquility, and yet nothing is more difficult to create. If the

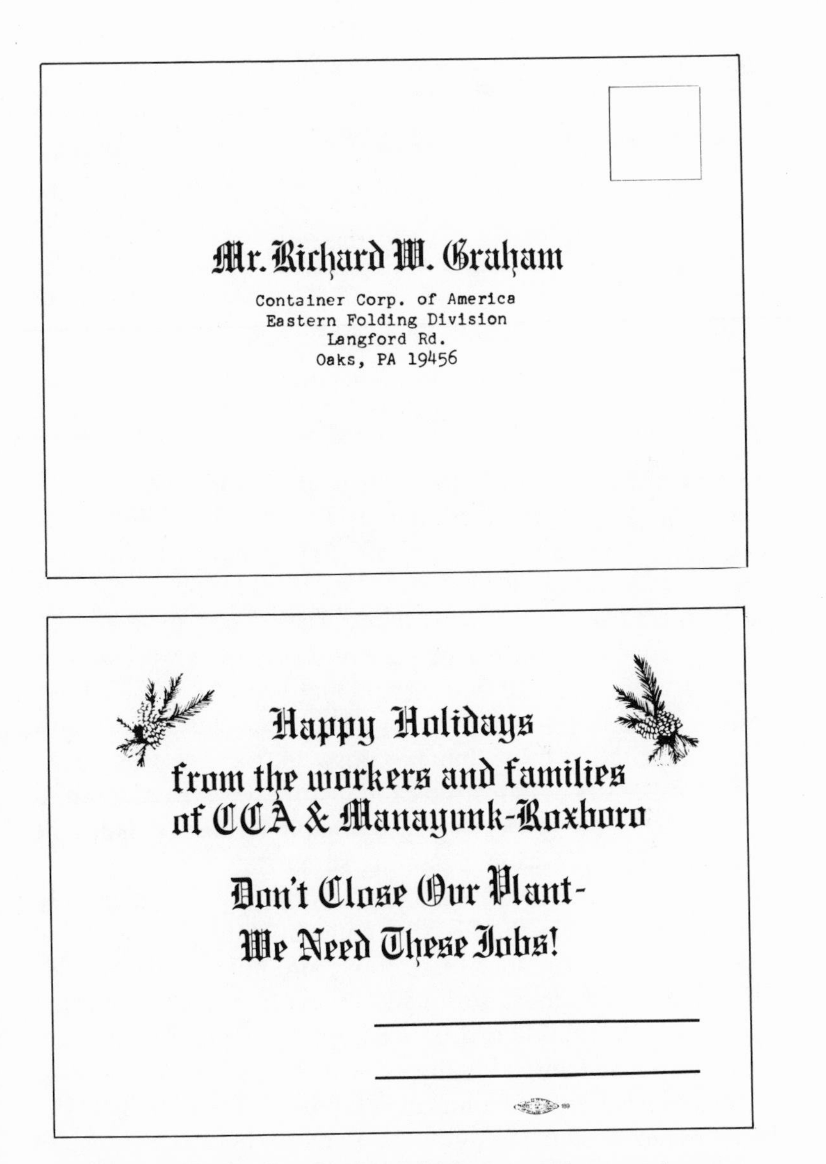

Christmas greeting sent by neighborhood residents to the manager presumed to be responsible for the closing of the CCA plant in Manayunk, December 1980.

> municipal bodies were made powerful and independent, it is feared that they would become too strong and expose the state to anarchy. Yet without power and independence, a town may contain good subjects, but it can have no active citizens. Another important fact is that the township of New England is constituted as to excite the warmest of human affections without arousing the ambitious passions of the heart of man.[6]

Thus, to de Tocqueville, "too much importance is attributed to legislation, too little to customs," in explaining why American democracy worked.

Unfortunately, most political leaders and writers today have forgotten these communitarian concerns of Jefferson, de Tocqueville, and early Americans in general. The irony is that a wide range of evidence is now affirming empirically what these traditional theorists could assert only instinctively. It now appears certain a strong, local community is essential to psychological well-being, personal growth, social order, and a sense of political efficacy. These conclusions are now being reached by more than just nostalgic radicals who are the strongest critics of industrial capitalism. They are emerging at the center of every social science discipline. In every field, the "community" position is gaining support.

I want, therefore, to elaborate on the ways in which well-organized neighborhoods still contribute to the values of personal freedom, social harmony, and political democracy that our society takes seriously. I want to argue as well that such neighborhoods can be an important asset to the economic life of a city. Finally, I want to outline a few basic principles behind economic development programs that take advantage of the strengths of neighborhoods, instead of merely taking advantage of the neighborhoods. There may be other grounds to defend viable neighborhoods against the encroachments of corporate development in this country. I would suggest, however, that these are the arguments that probably will make the most sense to the leaders of an otherwise liberal, secular society.

A wide range of evidence is now demonstrating the relationship between strong neighborhoods and personal stability. The most nearly complete study has been provided by Kai Erikson, who examined the reactions of the residents of Buffalo Creek, West Virginia, to a flood that destroyed their entire town. "In places like Buffalo Creek," Erikson maintained, "the community in general can be described as the locus for activities that are normally regarded as the exclusive property of individuals. It is the *community* that provides a context for intimacy, the *community* that represents morality and serves as the repository of old traditions."[7]

On this basis, Erikson notes that the devastation of a flood produced not only thousands of individual tragedies, but also a phenomenon that he refers to as collective trauma:

> The collective trauma works its way slowly and even insidiously into the awareness of those who suffer from it, so it does not have the quality of suddenness normally associated with "trauma." But it is a form of shock all the same, a gradual realization that the community no longer exists as an effective source of support and that an important part of the self has disappeared. As people begin to emerge hesitantly from the protective shells into which they have withdrawn, they learn that they are isolated and alone, wholly dependent upon their individual resources. "I" continue to exist, though damaged and maybe even permanently changed. "You" continue to exist, though distant and hard to relate to. But "we" no longer exist as a connected pair or as linked cells in a larger communal body.[8]

Federal relief officials were utterly oblivious to this aspect of the disaster, completely ignoring communal relationships in their reconstruction of living arrangements for the Buffalo

Creek residents. The results were poignant, from any point of view:

> Our own yard was a gathering place for the neighborhood children. There are children here, but we aren't even acquainted with them. There isn't one family in our trailer park that we were really close to. So we feel like we're in a strange land even though it is just a few miles up Buffalo Creek from where we lived.[9]

The organization of the community had become the focal point for an entire psychological orientation toward life. With its disruption, life itself fell apart.

This pattern was not unique to the rural setting of Buffalo Creek. One of the most famous studies of urban renewal in Boston characterized the reaction of residents of the West End to being forcibly removed from their neighborhood as, "Grieving for a Lost Home":

> Grieving for a lost home is evidently a widespread and serious social phenomenon following in the wake of urban dislocation. It is likely to increase social and psychological "pathology" in a limited number of instances; and it is also likely to create new opportunities for some, and to increase the rate of social mobility for others. For the greatest number, dislocation is unlikely to have either effect but does lead to intense personal suffering despite moderately successful adaptation to the total situation of relocation.[10]

Paul Levy found similar reactions among residents of Queen Village, Philadelphia, relocated to permit construction of Highway I-95:

> Kathy Conway:
>
> I just remember Fitzwater Street, Little Fitzwater, that would be east of The Bistro (Restaurant). Our house is on Strong and Fitzwater and it was the saddest thing. Each Saturday, you'd see

a new family going and it was really sad. It was terrible. That was most traumatic. You'd gradually each week hear of someone else who died, like Mrs. Domchowski had died, and each week like it was someone old and their daughter was maybe 58 years old and they would say, "Mom had a broken heart, like she just could not adjust to Ritner Street," or something like that. That was really the saddest thing. I can recall a good 20 people dying.[11]

The irony, finally, is that the sons and daughters of places like Buffalo Creek, The West End, and Queen Village are now themselves discovering the limitations of the autonomous existence which has replaced strong neighborhood communities in recent years. Daniel Yankelovich notes in his recent study of *The New Rules: Searching for Self-Fulfillment in a World Turned Upside Down* that the "search for community" has become a major phenomenon:

> In 1973, the "Search for Community" social trend, whose status my firm measures each year, stood at 32 percent, meaning that roughly one third of Americans felt an intense need to compensate for the impersonal and threatening aspects of modern life by seeking mutual identification with others based on close ethnic ties or shared interests, needs, backgrounds, age or values. By the beginning of the nineteen-eighties, the number of Americans deeply involved in the Search for Community has increased to 47 percent—to almost half of the population—a large and significant jump in a few short years.[12]

It appears that the dream of fulfillment in isolation upon which so much of modern society has been based has turned out to be as hollow as most early Americans assumed it would be.

Americans are rediscovering the relationship not only between community and personal stability, but also between community and education. To John Dewey, the father of modern liberal education, this relationship was clear. Dewey acknowledged that, "At present, the concentration of industry and division of labor have practically eliminated household and neighborhood occupations—at least for educational purposes," but with their loss he saw, "a real problem: how shall we . . . introduce into the school something representing the other side of life—occupations which exact personal responsibilities and which train the child in relation to the physical realities of life." His answer was the development of cooperative structures within the school itself:

> Where school work consists in simply learning lessons, mutual assistance, instead of being the most natural form of cooperation and association, becomes a clandestine effort to relieve one's neighbor of his proper duties. Where active work is going on, all this is changed. Helping others, instead of being a form of charity which impoverishes the recipient, is simply an aid in setting free the powers and furthering the impulse of the one helped. . . . To do this means to make each one of our schools an embryonic community life, active with types of occupations that reflect the life of the larger society and permeated throughout with the spirit of art, history, and science.[13]

This same sort of advice is offered to us today by Dr. Lawrence Cremin, one of our leading educational historians, and a disciple of Dewey:

> We live in an age that affirms individuality and pluralism, and, given what governments, including democratic governments, have done with their power in our time, we can understand and sympathize with the attractiveness of such affirmations. Yet, if Dewey taught us anything, it was that the public good is something more than the sum total of private goods, and that a viable community is more than a collection of groups, each occupying its own turf and each doing its own thing. Indeed, *Democracy and Education* is as much a work of social theory as it is of

> educational theory, and Dewey's own position is strikingly clear: There must be ample room in a democratic society for a healthy individualism and a healthy pluralism, but that individualism and pluralism must also partake of a continuing quest for community. In fact, individuality itself is only liberated and fully realized as the individual interacts with an ever widening variety of communities.[14]

Thus, despite our national obsession with "individualized" forms of teaching, liberal educators always have emphasized that without a community there cannot be a proper environment in which the individual can grow.

Strong neighborhoods are also being recognized as important allies in efforts to prevent crime. Dr. L. Harold de Wolfe, author of *Crime and Justice in America*, makes the point directly in a popularized version of his work entitled, *What Americans Should Do About Crime*: "The most important things Americans can and ought to do to reduce crime are steps to build and fortify a strong, cohesive, inclusive community. Such actions are mostly outside of the formal institutions of criminal justice. But if we take seriously the task of social defense, we must look first of all to the strength and soundness of the society we propose to defend."[15] Leaders of organized block clubs and citizen street patrols in cities throughout the country are now taking advice like Dr. de Wolfe's to heart.

Finally, political scientists are recognizing again the basic insight of Jefferson and de Tocqueville about political democracy in America: that it, too, depends upon a strong, local community. Sidney Verba and Norman Nie, two researchers from the University of Chicago, reached precisely this conclusion as part of a definitive study of participation in America in the late 1960's. Verba and Nie compared two theories concerning participation: one that says that we participate when we are given the wide range of information that a large city offers; another argues that we participate more heavily when we feel the sense of belonging which only a smaller community can provide. The survey results were clear: the smaller community was far more conducive to participation than impersonal cities.

> As cities grow in size and, more important, as they lose those characteristics of boundedness that distinguish the independent city from the suburb, participation declines. And it does so most strikingly for communal participation, a kind of participation particularly well attuned to deal with the variety of specific problems faced by groups of citizens. One last obvious point must be made here, for it has important implications. The communities that appear to foster participation—the small and relatively independent communities—are becoming rarer and rarer.[16]

Thus, whether we are talking about psychological development, public education, crime prevention, or political participation, a growing body of evidence developed by modern social scientists is reaffirming what classical and eighteenth-century theorists took for granted, namely, that a strong local community is essential to personal liberty and well being. On this basis, then, a city government that destroys such communities, for whatever reason, is wreaking havoc upon the very values that it wants to defend.

What about building a city's economy, however? Do urban neighborhoods make a contribution here, too? Or are they simply in the way? Clearly, all the social support in the world will not matter if everyone is out of work. Can the goals of economic development and neighborhood preservation be made compatible, or must city governments face an endless succession of agonizing choices between the two?

Obviously, if economic development means encouraging corporations to locate in urban areas on their own terms, then there is bound to be continuing conflict between development programs and neighborhoods. Yet if the real goal is creating jobs in the most effective way, then research suggests that neighborhoods can assume a greater significance in the process than they have had in years. The most recent evidence concerning job creation in cities reveals that it depends not on large corporations, but smaller firms, often located in or near urban neighborhoods. Moreover, many of the greatest problems facing these firms are those that relate to conditions in the surrounding community. When local governments view economic development in this way, then neighborhood improvement becomes an indispensable part of the process.

It was Jane Jacobs, writing in the *The Economy of Cities*, who first popularized a view of urban economic development that contradicted the notion that large corporations were singularly responsible for it. Jobs were created in cities, she argued, not principally by the arrival of new companies, but by the expansion of existing ones:

> Our remote ancestors did not expand their economies much by simply doing more of what they had already been doing: piling up more wild seeds and nuts, slaughtering more wild cattle and geese, making more spearheads, necklaces, burins and fires. They expanded their economies by adding new kinds of work. So do we. Innovating economies expand and develop. Economies that do not add new kinds of goods and services, but continue only to repeat old work, do not expand much nor do they, by definition, develop.[17]

Moreover, in cities, this process depended not upon established, well-developed firms, but on small companies with the potential to expand:

> But for a city to develop new work at a high rate means that its enterprises must have access to much inefficiently dispensed

> capital: many, many small loans and investments, a high proportion of them out of the routine, still other, relatively large, loans for swift expansion of goods or services that seem to be working out experimentally but which must go into large-scale production to become practicable—although it is not a certainty they will be.[18]

Thus, *The Economy of Cities* advanced the thesis that the nurturing of locally based small businesses, not the courtship of large, multinational corporations was the best way to proceed in developing the city as a whole.

Recent empirical analysis of job creation in the United States suggests that Jane Jacobs was right. In 1979, in a survey for the Economic Development Administration of over 5.6 million companies—encompassing 82 percent of private sector employment in the United States—Dr. David Birch of MIT came up with the following results:

1 Small firms (those with twenty or fewer employees) generated 66 percent of all new jobs generated in the U.S.
2 Small, independent firms generated 52 percent of the total.
3 Middle sized and large firms, on balance, provided relatively few new jobs.
4 There was considerable regional variation in this pattern. Small business generated all new jobs in the Northeast, an average percentage in the Midwest, and around 54 and 60 percent in the South and West respectively.

"It appears," Birch concluded, "that it is the smaller corporation, despite their higher failure rates, that are aggressively seeking out most new opportunities, while the larger ones are primarily redistributing their operations."[19]

This national finding fits the pattern in Philadelphia as well. Throughout the 1960's and '70's, while older manufacturing companies were moving away and local leaders were wondering what to do to replace them, the Philadelphia Industrial

Development Corporation (PIDC) was *saving* jobs by concentrating financial and technical assistance on hundreds of small firms that nobody else took seriously. By 1979, PIDC could issue the following report:

> This past year witnessed the achievement of a milestone by P.I.D.C. that cannot be matched by any other urban economic development agency—$1 billion in total projects assisted through P.I.D.C.'s financing programs. This level was attained through participation in over 1,300 transactions since 1958, involving 142,450 jobs—63,200 of them newly created jobs and 79,250 of them retained jobs.[20]

Clearly, the major trend in economic development in Philadelphia was taking place under everybody's noses.

It is this sort of economic development that not only coexists with neighborhood preservation, but in many ways also depends upon it. Four distinct relationships come to mind. Small businesses need entrepreneurial skills that community organizations can cultivate in their members. Neighborhoods also shape the work attitudes that employees bring to the job. The social conditions surrounding a plant will play a crucial role in determining whether it can survive; here, too, the environment of the neighborhood is critical. Finally, all of these factors—managerial ability, work attitudes, community stability—contribute to the willingness of lenders to provide capital to businesses. Each of these elements deserves careful consideration.

The notion that neighborhood associations can help citizens gain managerial skills is as old as De Tocqueville. As far back as the 1830's, he recognized the critical role that voluntary associations played in developing management skills:

> Men can embark in few civil partnerships without risking a portion of their possessions; this is the case with all manufactur-

> ing and trading companies. When men are as yet but little versed in the art of association and are not acquainted with its principal rules, they are afraid, when first they combine in this manner, of buying this experience dear. They therefore prefer depriving themselves of a powerful instrument of success to running the risks that attend the use of it. They are less reluctant, however, to join political associations, which appear to them to be without danger because they risk no money in them. But they cannot belong to those associations for any length of time without finding out how order is maintained among a large number of men and by what contrivance they are made to advance, harmoniously and methodically to the same object. Thus they learn to surrender their own will to that of all the rest and to make their own exertions subordinate to the common impulse, things which it is not less necessary to know in civil than in political associations. Political associations may therefore be considered as large free schools, where all the members of the community go to learn the general theory of association.[21]

Neighborhood activists are quite familiar with this process. The Southwest Germantown Community Development Corporation, for example, located two blocks from this writer's home, has produced not only several rehabilitated houses in the neighborhood, but also a number of trained real estate agents for local realtors as well. In marketing abandoned houses, these community workers gained the experience needed to pursue full-time careers in housing.

Community institutions also contribute to work attitudes in the neighborhood, especially among young people. Indeed, among the most effective youth development projects of the past two decades have been carefully supervised work programs in the community. Alternatively, among the most destructive programs have been the summer youth employment programs in cities where mayors have viewed them merely as a way of keeping kids off the streets. In 1974, the

Southwest Germantown Association grew so outraged at the lack of supervision provided to young people working in local recreation centers that it demanded and secured the right to run the entire program for itself. The indigenous parents' committee set up to administer the project developed far more stringent work rules than the city ever did. They understood the stakes, and they cared.

It goes without saying that a neighborhood can either contribute to or discourage crime and vandalism against the businesses in the community. Often, failure to involve neighborhood groups in local economic development efforts virtually insures that businesses will become targets. Nor is this a minor problem. A survey undertaken by the Philadelphia "Business Fortnight" revealed that of eighty firms that left the city over a five-year period, 28 percent cited "deteriorating neighborhoods and security problems in Philadelphia" as a primary reason. Only "better land sites and building features in suburbs" was a more common response. As a result, the city's Office of Housing and Community Development is now offering $100,000 for an anti-vandalism program on a major industrial corridor, involving neighborhood groups, manufacturers, commercial establishments, and city government itself. In this field, at least, the role of the neighborhood is now well understood.

Healthy neighborhoods, finally, can play a strong role in generating development capital for small businesses. Bankers and investors are aware that enterpreneurial skills, decent work attitudes, and local security are all products of the culture of the neighborhood and will rise or fall along with it. Moreover, given the high rate of small business failure, they will be exceedingly cautious in considering new enterprises for loans, especially if they are starting under precarious circumstances. Here, too, the existence of safe streets and strong community groups can go a long way toward persuading these lenders that the neighborhood is a stable place to do business.

Thus, if city governments want to help the sort of businesses that produce most of the new jobs in the United States—namely, small businesses—then they must make neighborhood preservation and development a priority as well. They should view neighborhood associations as centers for developing entrepreneurial skills, neighborhood job programs as the best way to promote decent work attitudes in young people, neighborhood citizen patrols as the best way to protect businesses and employees, as well as residents, from crime, and the broad range of neighborhood revitalization efforts as helping to persuade lenders to provide development capital for new enterprises. There is nothing "revolutionary" in any of this, either. Cities already are engaging in specific programs along these lines. Given the national obsession with downtown development and large-scale manufacturing, however, cities aren't giving these smaller efforts the attention they deserve. Yet, as planners are discovering, it is here where most of the new jobs in cities will come.

What, then, should city governments that want to coordinate efforts for neighborhood preservation and economic development do? How should they proceed?

Obviously, a number of specific programs must be designed, aimed at solving the mutual problems that we have been discussing—inadequate management skills, bad work attitudes, crime and vandalism, and the lack of development capital. Yet underlying these individual efforts must be a commitment to certain basic principles that will make neighborhood revitalization effective. Firstly, neighborhood leaders must become involved in the economic development planning process at all levels of government. Secondly, wherever possible, neighborhood consumers must have the option of buying locally produced goods and services over imported goods and services. Thirdly, and finally, neighborhood residents must receive the major share of jobs created by all city economic development efforts. We will conclude with a brief discussion of what each of these principles entails.

Neighborhood involvement in economic development planning is merely an extension of the growing involvement of community organizations in all areas of city life. Indeed, some cities and states are encouraging local groups to form their own democratically constituted community development corporations, with legal authority to acquire property and establish their own businesses. As part of its public commitment, for example, the Massachusetts Department of Commerce has published an extremely useful manual advising groups throughout the State on how to move in this direction. In Philadelphia, the city's Office of Housing and Community Development has funded the Institute for the Study of Civic Values to assist organizations in this area, including the creation of neighborhood credit unions, energy conservation cooperatives, and commercial strip development planning councils. In the past, cities felt that involving business, labor, and major civil rights organizations in economic planning committees was sufficient to represent everyone who cared. Given the direction of job development in the 1980's, neighborhood leaders must now be added to the list.

Trying to encourage as much local production of goods and services as is economically feasible both retains capital in the city and enlarges on the number of jobs that are produced to meet local needs. There is nothing new in this process. Jane Jacobs points out that it is a central feature of all rapidly expanding cities:

> The great cities of the world have had many repeated episodes of replacing imports and of explosive growth. Nobody knows when London had its first. It certainly had one in the thirteenth century (among other imports that London replaced at that time were the brass vessels it had previously been importing from Dinant, the city that overspecialized in its brass work and so had no other exports to make up for its losses when it lost that). But that was probably not London's first episode of import replacing and explosive growth. Paris, incidentially, was replacing from Dinant at about the same time as London. In the twelfth century, Paris had been no larger than half a dozen other French commercial and industrial centers, notable perhaps only for being less specialized in any way than the others. But in the thirteenth century Paris grew so rapidly that it became five or six times as large as any other French city, and this growth cannot be accounted for by any equivalent growth in exports.[22]

Some of the discussion of economic development in Philadelphia in the 1980's is moving in this direction as well. With a large number of hospitals in the city, planners are exploring the development of medical equipment manufacturing firms. A central goal of community-based energy weatherization programs locally is to stem the outflow of over $1.5 billion in capital each year to oil and gas producers elsewhere in the United States and the world. This sort of analysis must be undertaken on all areas of a city's economic life.

The most important commitment that a city government must make to its neighborhoods in economic development, however, is a fair share of the jobs to neighborhood residents. The proposition ought to be self-evident, but in the past it has not been. In Philadelphia, for example, 200,000 employees leave the city every night for the suburbs. This statistic, in general understood by the people of the city, has given downtown development a reputation of primarily benefitting suburbanites, not those whose tax dollars are paying for it. The

Green administration has responded to this criticism by including guarantees in the most recent Urban Development Action Grant for center city development that 30 percent of the contracts and 25 percent of the jobs will go to minority owned firms and workers. The Philadelphia Industrial Development Corporation has also produced a report showing that most of the jobs created by its efforts have gone to local, generally low-income residents of the inner city. In response, the Philadelphia Council of Neighborhood Organizations is now proposing the establishment of officially sanctioned neighborhood job coordinators who will receive information on all job openings in economic development programs, so that community groups can become a formal part of the placement process for the new jobs of the future. These connections must be made, if neighborhood involvement in economic development is to be meaningful.

The basic points, therefore, ought to be clear. Given what we know about the importance of strong communities in preserving a wide range of personal, social, and political values, urban leaders can no longer treat neighborhoods as if they were easily expendable when other objectives are at stake. To the extent that cities need to produce new jobs, the evidence suggests that small businesses, often located in neighborhoods, will be the greatest contributors, not large corporations that relocate from the outside. Thus, city governments should work to coordinate programs between local businesses and neighborhood groups that emphasize management development, job training, crime prevention, and the search for new sources of capital. In conducting these programs, officials need to involve neighborhood groups on all committees, encourage indigenous production of locally used goods wherever possible, and guarantee that a fair share of all new jobs go to the neighborhood residents themselves. This approach might not be exactly what's good for General Motors. It will, however, be good for our cities, and for the wide range of people who still live in them.

Notes

1. Mayor Coleman Young, "What's Good for General Motors," *CBS Reports*, July, 1981.
2. Ibid.
3. Thomas Jefferson, quoted in Gary Wills, *Inventing America: Jefferson's Declaration of Independence* (New York: Doubleday, 1978), p. 205.
4. Thomas Jefferson to Samuel Kerschevalu, Monticello, Sept. 5, 1816, in Edward Dumbauld, ed., *The Political Writings of Thomas Jefferson* (Indianapolis: Bobbs-Merrill, 1955), pp. 97–98.
5. Alexis de Tocqueville, *Democracy in America* (New York: Vintage, 1945), p. 334.
6. Ibid., p. 68.
7. Kai Erikson, *Everything in Its Path* (New York: Simon and Schuster, 1976), p. 154.
8. Ibid., p. 154.
9. Ibid., p. 211.
10. Marc Fried, "Grieving for a Lost Home," in James Q. Wilson, *Urban Renewal: The Record and the Controversy* (Cambridge: MIT Press, 1967), p. 377.
11. Paul Levy, *Queen Village: The Eclipse of Community* (Philadelphia: Institute for the Study of Civic Values, 1978), p. 59.
12. Daniel Yankelovich, *New Rules: The Search for Fulfillment in a World Turned Upside Down* (New York: Random House, 1981), p. 251.
13. John Dewey, "The School and Society," from Reginald de Archambault, ed., *John Dewey on Education* (Chicago: University of Chicago Press, 1964), pp. 301, 310.
14. Lawrence Cremin, *Public Education* (New York: Basic Books, 1976), p. 72.

15. Harold de Wolfe, *What Americans Should Do About Crime* (New York: Harper & Row, 1976), p. 109.

16. Sidney Verba and Norman Nie, *Participation in America* (New York: Harper & Row, 1972), p. 247.

17. Jane Jacobs, *The Economy of Cities* (New York: Random House, 1969), p. 49.

18. Ibid., p. 100.

19. David Birch, "The Job Generation Process," Cambridge, MIT Program on Neighborhood and Regional Change, 1979, p. 8.

20. *PIDC Annual Report: 1979* (Philadelphia: PIDC, 1979), p. 2.

21. De Tocqueville, *Democracy in America*, vol. 2, p. 125.

22. Jane Jacobs, *The Economy of Cities*, pp. 157–58.

Chapter 13

Economics and the Justification of Sorrows

John C. Raines

Economics is the study of scarcity. If there were no scarcity, there would be no need to understand the mechanisms by which costs are allocated, and hence no need for economics. It has been called "the dismal science" because economics concerns itself with the justification of sorrows, with *limits* to satisfactions.

Such justifications depend, in turn, upon how claims to social benefits are perceived. "It is not necessary," Emil Durkheim has pointed out, "that a person have any more or less for social tranquility to prevail; it is only necessary that they *feel they have no right* to any more or less."[1] What we believe we have a right to, that's what's crucial.

It is crucial because all economic practice—whether traditionalist or bourgeois, capitalist or radical—depends upon a society's common agreements concerning what, out of the total of limited social benefits, each party has a right to claim for itself. Without this sense of legitimacy, a society either unravels into unending and unnegotiable rival claims, or is forced into the drastic inefficiencies of a police state.

We may take as an example of this struggle over legitimacy the present flight of jobs and of capital from our country's

Frostbelt. How are we to count the overall costs and benefits of all this relocation? How are we to weigh the loss of a job against the promise of the more productive use of investment capital elsewhere? What claims are to be admitted into the making of this calculation?

What is the cost to society, for example, of a working-class father who cannot find a place for his son in the factory? What is the cost when he loses his own place there? How are we to calculate the claims of a mother who has to go to work when her children are still young and wonders what will happen now that she's tired all the time? How are we to estimate the costs to a nation when the way many people have of giving gifts to the next generation, and so making sense out of sacrifice, begins to unravel?

Philadelphia, for example, lost over 140,000 manufacturing jobs in the past ten years. The same thing was happening in Baltimore and New York, in Cleveland and Chicago, in the whole of the older industrial region of our nation. For every hundred jobs lost in the Frostbelt, only eighty-five new jobs were created; while in the Sun Belt, one hundred eleven new jobs replaced every hundred lost.[2]

Many economists reply that working-class people of the North need obviously to move. President Carter's "Commission on the Agenda of the '80's" made this point quite bluntly. It suggested that the federal government shift "from a place to a people oriented policy," and no longer try to keep jobs in the industrial cities of the North but help individual workers retrain for new and more efficient work in a new location.[3]

But such advice simply ignores the human costs to working people of having to move in order to find work. Why? Because many have meanings that do not travel well, like loyalty to extended family or attachment to neighborhood and old friends. These are nonpecuniary values which do not show up on econometric readouts. But they are real human values; and they are values that can and do get injured. Prices are paid; only at present these prices do not get factored into decisions about the "most efficient use" of reinvestment capital. The result is an inaccurate picture of how social costs and benefits in fact get allocated in our society. The counting system is biased in favor of unrestricted capital mobility, without regard to the costs this mobility may extract from family and community stability.

Still, many would claim that economics cannot get into the business of counting noneconomic values. How can we quantify, for example, or calculate in economic terms the cost of leaving parents or grandparents behind in a move from the Frostbelt to the Sunbelt to find work? Rather, the task of those who manage the market, some say, is to make accurate economic decisions on the basis of productivity and consumer demand, so that reinvestment capital may flow where it can be used most efficiently. In the end, they conclude, only an efficient economy is able to compete, and from its success everyone will eventually benefit. In the language of the professional economist, this is an argument for Pareto optimality: that an economic policy is "Pareto efficient" and should be adopted if, as a consequence, everyone is better off, or at least on average people are not worse off.

Notice that this justification, which focuses exclusively on pecuniary benefits ("worse off" means economically worse off), is, nevertheless, a form of ethical argument. Its intention is finally moral justification. Its type of moral reasoning is utilitarian: to seek the greatest good for the largest number. Economics, even when it considers itself purely scientific and technical, is based in the end upon ethical claims. It must justify its allocation of sorrows.

Why then do economics and ethics so often get widely separated? Why can a well-meaning businessmen say "economic decisions have nothing to do with morality?"

To find the reason for this separation of economics from ethics, we may look to the founding father of modern economics, Adam Smith. Ironically, it began when Smith, a professor of moral philosophy, sought to console himself in face of an ethical dilemma: how to relate *positively* the given conditions of mass poverty to the evident lack of dependability of personal benevolence beyond the narrow sphere of a few close associates. Smith wrote *Inquiry into the Nature and Causes of the Wealth of Nations*, according to his own testimony, as his personal testament of hope for "the poor of London." The free market which he argued for in that book was for him a theodicy, the reconciliation of present misery with eventual benefit and relief, and a reconciliation based not upon human generosity but, of all things, upon human greed.[4]

Smith wrote some of the most influential lines in all economics:

> Every individual is continually exerting himself to find the most advantageous employment for whatever capital he can command. It is his own advantage, indeed, and not that of society, which he has in view. But the study of his own advantage naturally, or rather, necessarily, leads him to prefer that employment which is most advantageous to the society. . . . In this case, as in many other cases, he is led by an invisible hand to promote an end which was no part of his intention.[5]

What Smith is arguing for is unhindered capital mobility. The central hinge of his argument is *moral paradox*: that the pursuit of private gain will lead "necessarily"—i.e., without conscious human intent—not to what we might expect, for example the empowerment of greed, but to public good. Moral paradox makes the free market palatable. Unintended, the market transforms private vice into public virtue. Moral intention need have no part, nor need moral issues consciously be considered, since the free market operates automatically for benign effects.

How?

As Smith argued, the free market associates consumers, who rationally pursue their own self-interest, with competing producers. The consumer seeks the highest quality product at the lowest cost. If the manufacture and distribution of a product proves profitable, it can be only because its benefits to buyers exceed the costs of production. The cost mechanism thus measures the value of other possible products to which that same labor and capital might have been put, and channels resources into more productive uses and away from less productive ones. Moreover, among competing producers only growing efficiencies can maintain product quality at comparative prices. In the search for expanding efficiencies, producers seek the efficiency of scale—larger and larger units of production. This will increase the need for labor which will, in turn, increase not only the total number of potential consumers but their ability to develop their wants beyond subsistence. Thus, the overall economy grows. The total amount of goods and services expands. And all of this increase stems from the freedom of entrepreneurs to move capital where they wish, when they wish.

True, some may get more of this bounty and others less. But everyone gets more than they had in the beginning. Thus, to relieve "the poor of London" it was only necessary, Smith argued, to unleash the dynamism inherent in a free market from what was in his day the combined restraints of the king's

Frank Maguire is a peacemaker.

He believes in the peace that follows a hard day's work. He believes in the peace of mind that comes from knowing that you have a job. In Philadelphia's blue-collar neighborhoods, Frank Maguire — the peacemaker — fights to keep jobs secure.

The Campaign for Human Development supports peacemakers like Frank Maguire in his work with the Philadelphia Unemployment Project (PUP). Funded in part by the Campaign, this self-help group was the catalyst in the development of the Delaware Valley Coalition for Jobs. It has organized service centers for the unemployed, published research on the problems of the unemployed, and spearheaded efforts on behalf of the jobless.

Most important, perhaps, PUP is working to stem the tide of factory closings in the Philadelphia area, securing the jobs of thousands of unskilled workers.

Other CHD-funded projects tackle tough problems in the areas of housing, health, legal aid, education and communications. Your contribution to the Campaign helps hundreds of self-help groups bring people out of poverty.

Join the peacemakers.

CAMPAIGN FOR
HUMAN DEVELOPMENT
UNITED STATES CATHOLIC
CONFERENCE

STOP
RUNAWAY
SHOPS

IF
YOU
WANT
PEACE,
WORK
FOR
JUSTICE
PAUL 6

Campaign for Human Development, U.S. Catholic Conference—used with permission

Campaign poster: Campaign for Human Development.

ministers, whose chief dedication was to the size of the royal treasury, and the multitude of traditional workplace rules involving guilds and inherited claims to jobs.

Note that all this social benefit is deeply rooted, paradoxically, in individual self-pursuit. For heaven, man may need the grace of salvation; but for easing the burdens of this life, such intervention can be done without: indeed, can be done *better* without.

Lord Keynes, who otherwise had many differences with his distant predecessor, nevertheless, on this key point remained in perfect accord. In 1930, in the middle of the Great Depression, he speculated about a time, perhaps for his grandchildren, when everybody would be rich. "But beware!" Keynes warned.

> The time for all this is not yet. For at least another hundred years we must pretend to ourselves and to everyone that fair is foul and foul is fair; for foul is useful and fair is not. Avarice and usury and precaution must be our gods for a little longer still. For only they can lead us out of the tunnel of economic necessity into daylight.[6]

Keynes shared not at all Smith's passion for un unregulated market. But he did share his predecessor's theodicy—the moral paradox which sets economics at such remote distance from the, to them, disruptive voices of moral claim.

Why disruptive? Because what person or group of persons can so transcend their own self-interest as to attain an unbiased moral perspective? The market permits a beneficial moral modesty that trusts people acting in pursuit of their own interest far more than avowals of disinterested benevolence or claims to objective moral insight.

This argument did and does fall with especially persuasive weight upon those attuned to the pervasive selfishness operating between human groups, not excluding those groups which claim not to act selfishly. Moreover, it is an argument whose roots go far back into Christian thought with its sometimes heavy emphasis upon human sin. By Adam Smith's day, Deism had drained the ideas of Divine Providence and Grace of much of their power to console. But a theodicy in decline found a successor for many in the idea of the free market and its beneficent, though wholly unintended, social results.[7] For many of Smith's contemporaries, as for many of his followers in the discipline of economics today (and some who "follow" only at considerable ideological distance), it was and is a theodicy adequate for the justification of human sorrows. For it removes, supposedly benignly, the social disruption of unresolvable claims to superior moral right—unresolvable because reasoning is itself poisoned by the bias of its own perspective.

Free market ideology rests upon this paradoxical justification: economics need seek—indeed must seek—no higher calling in order to produce its beneficent results.

By thus removing moral considerations from marketplace decisions, and reconciling, it was hoped, all participants in the market to the beneficial effects of this removal, decisions as to the use of investment capital became, so it seemed, simply technical. Noneconomic factors bore no relevance.

And thus we arrive at our present situation where economics, the study of the allocation of costs, does not count the costs of capital mobility upon family and community stability. The theodicy of moral paradox need tally only individual greed, not human connectedness. The danger, of course, is that such a perspective underestimates precisely that which it refuses to tally: namely, the degree to which humans are social beings, and not just individual.

This danger becomes evident in two ways. First, the free market perspective tends to ignore the possibility that many goods people seek are good only as shared, as a *common good*.[8] Second, it tends to underestimate the degree to which human behavior (including selfish behavior) is not a fixed entity. Social structures which tolerate the unrestricted exercise of personal greed may well encourage the expansion of self-preoccupation and self-pursuit. This is not to say that people are not inherently selfish; only that the balance of selfishness and sociality is not fixed for all times and may be modified to a degree by what individual attributes a given society admires and rewards.

Moreover, the consolation of moral paradox, looking as it does to the unintended transformation of private vice into public benefit removes, or seems to remove, the burden of conscious moral choice. It removes, prematurely, the need for common discussion and reaching of common agreements about those things which are good only as we share them in society, not as individually we own and use them. Environmental pollution, for example, or the unequal distribution of economic benefits and its implications for the democratic conduct of power are clearly *not* managed benignly by the market.

The emphasis one places upon the individual versus the social, or the private versus the common good, is an old debate in western civilization; although today many do not bother to debate it, and operate uncritically upon the assumption of the pervasiveness and fixed quality of human greed.

The bias of this essay is towards the reassertion of the social and communal aspects of human life. For it seems evident to this writer that *selfishness presupposes, and in fact depends upon, sociality*; that the free market rests upon social foundations it cannot provide for itself or by itself sustain. As I shall argue in detail later, there is simply no entity for the word "self" to refer to without presupposing sociality; and, more obvious, even selfishness necessarily entails shared social agreements—namely, about what it is desirable individually to pursue and possess.

If we are to make any headway in bringing ethics and economics together beyond this bias towards individualism and the theodicy of moral paradox, then we must go back to where the path first begins to divide, to where self-interest becomes assigned to *this* world and sociality only or primarily to the next world.[9]

Contrasting Christian Views of Society

"There is nothing so social by nature, so unsocial by its corruption, as this race."[10] With these words St. Augustine summarizes one of the two major strands of reflection upon selfhood and society developed by Christianity in the West. It was to have the greater effect because it influenced not only subsequent Catholic thinking but also the Protestantism of Martin Luther, and, especially, of John Calvin.

The other tradition stretches back to Thomas Aquinas, and behind him to Aristotle and the classical idea of society as a *res publica*. It emphasizes the continuing sociality of humans, even after the Fall, and has had its major influence upon that feudal culture which was destined to be overwhelmed by the rising bourgeoisie, whose merchants sought economic freedom from traditional feudal restraints.

This division within Christian thinking about selfhood and society—one stressing selfishness, the other sociality—was crucial for the development of free market theory. Not because there was any direct dependence by economists upon the thought of the Church, but because of a prevailing ethos in Protestant England and America which viewed human behavior as, for the most part, individually and selfishly motivated. As Max Weber has noted,

> Combined with the harsh doctrine of the absolute transcendentality of God and the corruption of everything pertaining to the flesh, this inner isolation of the individual . . . forms one of the roots of that disillusioned and pessimistically inclined individualism which can even today be identified in the national character and the institutions of the peoples with a Puritan past."[11]

Now it is not my claim that Christian reflection upon the nature of society is the only or even the most important influence upon the development of free market ideology. I maintain only that these Christian ideas had an influence, and that without these ideas public discussion in the West about the nature of human community would have developed quite differently. Thus, it is not too much to say that it was upon the distant platform of Augustine's pessimism about man's earthly possibilities that later free market ideologues were to build their justifications of human sorrow.

In sharp contrast to Augustine, selfishness to Aquinas presupposes sociality. The self's loving itself, even inordinately, rests upon a platform of commonalities. As Aquinas puts it,

> It is natural for man to be a political and social animal, even more so than all other animals, as the very needs of nature indicate. For all other animals nature has prepared food, hair for covering, teeth, horns, claws as means of defense, or at least speed of flight. Man, on the other hand, was created without any natural provision of these things. But, instead of them all, he was endowed with reason, by the use of which he could procure all these things for himself by the work of his hands. But one man alone is not able to procure them for himself; for one man could not

sufficiently provide for life, unassisted. It is, therefore, natural that man should live in the company of his fellows.[12]

Self-interest, for Aquinas, if perceived accurately, carries us not to individual selfishness but to "the common good." It is what we have in common that blesses us: "Society is the moral and stable union of a number of men who are striving by their united effort for the attainment of a common goal."[13] The just society, which is also the rational society, is constituted in this "common good." And to this common good, the private good of the self cannot be fundamentally opposed except by way of self-contradiction. To be rational, including reasoning about one's own best interest, is to pursue the health of the whole.

Had this become the dominant thinking on selfhood and society in the Christian West, the moral defense of the free market would have been far more difficult. But what happened instead was that the Protestant Reformation and—of especial importance for the English-speaking world—John Calvin, were far more influenced by Augustine than by Aquinas. It was Augustine's view of sin and its devastating effects upon reasoning and the moral conscience which grasped the imagination of the Reformers.

For Calvin there is no natural, rational, or moral faculty except one radically tainted by the Fall. This is what he meant by his doctrine of "total depravity": *not* that our rational and moral capacities are nothing but depraved, but that all our capacities, including reason, are involved in the ruin of sin. That is why Calvin left behind his humanist studies (his first book was on the Stoic moral idealism of Seneca) and turned his attention instead to the Revelation of Divine Truth in the Bible, in which alone is to be found, he thought, the only "sure and certain guide" to personal rectitude and corporate order. Precisely because of the pervasiveness of selfishness, the social order, or rather, disorder, requires for its correction this Grace of Revelation.

For those less confident of such Divine Intervention, yet who with Calvin and Augustine viewed the human predicament with profound moral suspicion, the problem of social order now became acute. Indeed, in the eighteenth and nineteenth centuries in the West, the world did seem less and less God-dominated and more and more the responsibility of mankind, however unsuited humans might be for that task. Precisely this turmoil of lost hope provided the opening wedge for the appeal of Adam Smith's promise of the free market.[14]

Thomas Hobbes argued in his *Leviathan* that it is because "each is at war against all" in our natural state that we humans contract together under a central and powerful authority (Hobbes preferred absolute monarchy) in whose will is found both the plan of social order and the power to impose it. But Smith's vision provided a more benign resolution. Human selfishness, including the self-interested reasoning of individuals, could be associated by the mechanisms of the free market to a less constrained social benefit. For the mailed fist of the monarch could be substituted the invisible hand of the market. Not in autocracy, nor in the illusory common search for the common good, but in the socially unrestricted pursuit of personal advantage, one can find hope, a hope with the added advantage of being established, for Smith, upon a sober and realistic grasp of the nature of human group behavior.[15]

But how realistic is this realism? Does the key of selfishness unlock all the important doors to understanding human behavior? Or does it ignore the necessary place and claim of society?

The Primacy of the Social

It seems evident to me that without society there is no self to be selfish about. At least this claim was the crucial intellectual contribution of G. H. Mead and the American School of social psychology he represents.

That the social is primary in regard to the human has become by now less a claim than a taken-for-granted starting point in most American sociology and anthropology. But it is a starting point that has little penetrated American political and economic thought, which, for the most part, reflect upon human motivations in terms of the idea of "interests" (self or group).

The primacy of the social—what does this mean?[16]

Mead and others have pointed out that human behavior is not first of all instinctual; rather it is behavior that is modeled after shared meanings. "What is astonishing," says anthropologist Clifford Geertz, "is not how much we humans can learn but how much we have to learn in order to survive." For example, the infant does not begin by knowing what a cry means. The infant lives in a kind of chaos of random and ad hoc gesturing. It learns what a cry "means"—namely an announcement of distress and an appeal for succor—by way of observing the interpreting response of the attending adult. The adult responding with concerned attention and active comfort assigns the meaning to the infant's gesture. This meaning once learned, the infant can then cry *with intention*, intending a specific and predictable response. The infant is no longer trapped, so to speak, inside its idiosyncracy. It has entered the world of reciprocally designated meanings.

Starting from these relatively simple gestures of crying and laughing, the infant builds up common recipes of behavior.

Language and the careful profiling of eye and voice and body gesturing into public performances which communicate intentions rather than simply astonish—all this is *learned* behavior, not preprogrammed by instincts. Human behavior, as Geertz points out, is more like a wink than a twitch.[17] It is behavior which intends meanings. And meaning is a joint performance; it is something we learn to do and continue to do only together.

It is as an intentional self—one who unfolds distances from her immediacy to the world, makes intentions toward that world and seeks predictable responses there—that we begin to experience ourselves as a "self." We begin to notice ourselves as the orchestrators of our effects upon the world, although those behaviors, as bearing steady meanings, had to belong to others before they could belong to us.

The self as a center of intentionality, as a coherent and continuous identity, as that which attains a steady focus in the public arena of attention—this "self" is fundamentally social. Without this sociality the self cannot intend anything, including its own self-interest.

What is true at the beginning of each human life continues throughout the adult years. It is not that we are not selfish. But the general health, the trustworthiness of what we have in common, is presupposed in any merely self-interested activity. For it is what we have in common that allows us to take up a routine, taken-for-granted posture towards others. Before we can exploit others for our own selfish ends, we must trust the world we have in common with them enough to overcome our being swamped by fear and thus growing unable to take up any practical attitude. Paranoiacs are notoriously poor at achieving self-interested goals. They are so filled with suspicion that they cannot attain the distance and perspective required for cunning or calculation.

Both selfhood and selfishness depend upon sociality—the common good of a shared and trusted public environment. And the continuing health of that public environment depends, in turn, upon what I shall call the *everyday virtues of the common life.*[18]

I mean by this term such things as the necessary assumption of relative dependability in each other's word and work that allows us to go forward with our everyday and practical concerns without constantly tripping over our suspicions of one another. We really don't bother to question whether the bus driver is a mass kidnapper in disguise when we get on the bus. We can't be bothered to! Not if our practical life with others is to get on with its everyday business. We depend upon, because we must take-for-granted, the trustworthiness of others in our everyday world.

Or again, there is that virtue, so inordinately and uncalculatingly practiced most of the time, of giving the other guy the benefit of a doubt. This is the graciousness of spirit which fuels the common life even as it opens that life possibly to be expropriated by the unscrupulous. But even the unscrupulous depend upon that which they steal from. Without this astonishing (when we stop to think about it) generosity of everyday living, with its manifold gestures of uncoerced and unrewarded helpfulness, a society begins to narrow and becomes inexhaustibly calculating and envious, indeed comes at last to a kind of standstill. Society is an ongoing display of common grace: that when I trip getting on the subway car, someone's hand, without thinking, reaches out to keep me from falling. Compared to our dependence upon this everyday generosity of the common life, claims to independence and self-reliance are like Chanticleer's, thinking his crowing brings up the sun.

What I am pointing to here is the whole thickness and matrix of the shared and everyday world, the simple truth of our common exposure before life and consequent ineluctable reliance upon one another. This is the fundamental order of our life together, and to it all self-pursuits, all market negotiations, remain strictly second-order realities—able to disrupt but not able to create or to survive without.

Both in our infancy and in our maturity we humans are more social than selfish. We remain more dependent upon the common good than upon our private possessions, whose ultimate value to us deteriorates in precise relationship to the deterioration of the common life.

Today in America, it is precisely this moral quality of our life together, our sense of shared belonging, that is unraveling. The rapid mobility of capital investment that does not include within its calculation of costs the social cost to community is undermining our nation's communal foundations.

How is this happening? And what should we do about it?

Capital Mobility and Social Costs

We began by asking what the cost to our nation is of a father who cannot find a place for his son in the factory, or who loses his own place there. We may ask further, what does a nation lose when it loses a community—because its factories close—a community which for three or four generations of a family has nurtured them with tested and known teachers, store-keepers with their moral uses of gossip, with the continuity of lifetime friends, and with the common project of decent streets enacted by neighborhoods of adults knowing which children's behavior to report to whom? What are the social costs of more or less requiring extended families to break up as various members head in different directions to find work, or at least to find decent paying work?

Contemplating the closing of a plant raises a whole myriad of cost and efficiency calculations, only part of which (and perhaps the least consequential part) is now in fact tallied. Only recently have we begun to gather evidence on community costs, and it shows that the costs of a plant closing can be quite substantial: to families, to neighborhoods, and to the wider region. But the proper way to factor these costs into decisions as to the most efficient social, as well as economic, use of capital investment is a discussion that has hardly been begun. Many in our society continue to insist that it is unnecessary. Yet the evidence mounts which indicates that heavy damage is being done to the moral foundations of our nation by capital decisions that ignore community.

What is this evidence?

Evidence is widespread that job loss is positively correlated to increased rates of alcoholism,[19] of wife and child abuse,[20] increases in suicide,[21] and in homicide,[22] and a rising incidence of heart attack and other symptoms of acute physical and mental distress,[23] including first time admissions to mental hospitals and prisons.[24]

The most common response to job loss is decreased self-esteem and generalized depression and anger.[25] This is reflected in the words of a twenty-four-year-old, working-class male from Philadelphia who lost his job. "Unemployment gets you scared," he said:

> You start studying yourself. Are you a person? Are you able to work? I used to start looking at the papers, and sometimes it would only be one sheet of jobs in the paper, in the whole paper, one little sheet. There was nothing you could do. Sundays come and you look, and there's nothing you can do. It would be hundreds of jobs there, and there's nothing you can do.
>
> We both were on welfare at one time—Charlotte and me. [Charlotte is his live-in girl friend.] That's the most depressing

> part of our life. We were both collectiving DPA. All the money went to pay the bills. We had no money; we didn't have no food; it was really hard. We went through a really bad thing mentally. When you're living on spaghetti and peanut butter, how can you get friendly? You have to take it out on each other. We were buying our clothes at a thrift shop. We're talking about 25 cents for pants; 75 cents for shoes and stuff. This is America?[26]

Beyond such individual dislocation, job loss leads to lost confidence and lost hope *within families.* A wife of a dockworker living in South Philadelphia told me of the sadness which weighs upon a father when he has to tell his son that he can't expect to follow his dad onto the docks, that because of containerization there won't be enough jobs. "For middle-class people," she said, "hope means change; but for working people hope is when things stay the same." The continuity of fathers finding a place in the work world for their children, of the extended family of uncles and cousins sheltering and ushering the teenage passage into adult life—all this finds itself failing, and in that failure, humiliated. It's as if a whole world is falling apart. A way people had of making sense of their lives is shattered. And the emotional response often corresponds closely to the classical grief syndrome: denial, anger, guilt, depression, emotional confusion, and lethargy.[27]

Feelings of impotence, of not being able to protect one's own, a gnawing sense of incompetence are some of the mental costs of job loss. They are prices paid not just by individuals but by whole families. They are injuries which come down *between* generations, clouding the younger generation's picture of its own adult future. It hinders respect, denying the ability of one generation to give effective gifts to the next.

Dr. Harvey Brenner, of Johns Hopkins University, using sophisticated statistical techniques, projected, in terms of its mass effect, the mental and physical health results of job loss.

For every one percent increase in sustained unemployment, Brenner's figures project 37,000 additional deaths, including 920 more suicides, 650 more homicides and 500 more deaths due to cirrhosis of the liver. Moreover, for every one percent increase in sustained unemployment there will also be 7,000 additional admissions to state mental hospitals and 3,300 additional state prison confinements.[28]

Besides these health impacts, job loss has a host of economic effects upon individuals and communities. The same twenty-four-year-old quoted earlier spoke about the anguish he saw in the unemployment line.

> Seeing that unemployment office for the first time is probably the most horrifying thing I seen in my life. When I say horrifying, not like scary, but just seeing people being agonized. You're in line for three and a half hours. You talk about money, earning money. Unemployment is too low. I seen grown men that been working at the same job 15 years cry in the unemployment office. Because they are afraid of losing their house, or their marriage is breaking up.

The Youngstown Sheet and Tube plant closed in January 1978, and 4,100 workers lost their jobs. Two years later *Fortune* magazine did a follow up on the workers. Of the original 4,100, 35 percent had been forced into unwanted early retirement with an income half what it had been when they worked. Another 15 percent were still unemployed two years later; while 20 to 40 percent had taken drastic pay cuts in finding new work.[29]

Some have suggested that the hope of our older Frostbelt cities is that the departing industrial jobs be replaced with jobs in the service area: in banking or law or the tourist industry. The latter has led to the present curiosity of twenty-six American cities, each vying with the others to become an "international city." This vision of a viable Frostbelt future, however, is hardly viable for factory workers, since statistics show that the average factory wage in Pennsylvania in December 1979, for example, was $285 per week, while the average service wage was only $170. The whole idea of a "service" city makes little economic sense either to blue-collar workers facing pay cuts or to the metropolitan regions that must endure this diminution of average worker purchasing capacity.

Prices paid for factory shut downs do not stop with families. There are economic costs to the city and the region. Closing plants mean a shrinking tax base, which affects school budgets, the quality of police and fire protection, municipal support for libraries and museums, for parks and playgrounds and summer job programs. The whole infrastructure of civilized life together is eroded. Local communities are caught in the double bind of the loss of tax revenue combined with rapidly escalating welfare and social service costs. There are more needs and fewer tax dollars to meet them.

Moreover, there are hidden costs, such as increased street violence, or sharply higher school dropout rates among families of the chronically unemployed, and a higher rate of divorce and family abandonment. There are more mortgage failures. And there are fewer loan applications for home improvements. Neighborhoods go down hill. In Kensington, a Philadelphia blue-collar neighborhood, a young widowed mother of five remembered how growing up "on these same streets, summers we was always barefoot. Then they wasn't full of glass like now."

There are other more deeply hidden costs. There is the ripple effect of a plant closing on other business in the area. Or there is the underutilization of the industrial substructure of water mains and sewers and heavy-duty electric power lines operating at far less than capacity, of miles of rail siding that now trail uselessly from abandoned factories, and half-full public busses and trolleys which once carried hundreds of thousands to work in a fuel-efficient manner.

Another cost of capital transfer and plant closings—and one that is almost never talked about—is its effect upon managers and their perception of their careers. In a suggestive article entitled "Managing Our Way to Economic Decline," Professors Robert Hayes and William Abernathy of Harvard Business School point out the unintended result of rapid managerial turnover, especially in the junior ranks. It forces junior executives, they concluded, always to maximize short-term bottom line figures.[30] "I know—we all know—of people all over this town," says David Joys of the Russell Reynolds Consulting Firm of New York City,

> who are running their companies into the ground, taking huge, quick profits and leaving them a shell. And when you look at their contracts, it's easy to see why. What does it matter to them what happens 10 years from now? They're building giant personal fortunes, and *appear* to be running their companies terrifically, and in 10 years, when there's nothing left, they'll be long gone.[31]

Something like this seems to have happened in Philadelphia at the Eaton, Yale and Towne Corporation where 600 employees, some with thirty years experience, face the shutdown of their plant. Ever since a conglomerate bought the local plant in the mid-'60's, portions of the manufacturing operation, which makes fork-lift trucks, have been moved south or overseas. First service parts assembly, then transmission cylinders—bit by bit workers saw their jobs transferred elsewhere. As the plant in Philadelphia fell into obsolescence, owners used tax incentives for depreciation to build new plants elsewhere. Slowly, what was once a profitable enterprise was milked dry, improving after-tax bottom line corporate profits, while leaving behind a highly trained work force forced out of work.

The irony of this situation, as Hayes and Abernathy of Harvard Business School point out, is that American business today suffers from a shortage of precisely the kind of *stay-put, production-trained managers* who possess wisdom that can be gained only over long-term, direct involvement with the production process, and an intimacy with experienced workers and their indispensable feedback.

Preoccupation with short-term bottom line figures works against long-term productivity. Shoving plants and jobs rapidly around the country, or between countries, deteriorates not just worker families and their neighborhoods. It poisons the manager/worker relationship upon which rests long-term productive efficiency.

Few of these costs are presently registered as costs by the free market. Quite the opposite, the unregulated velocity of capital mobility is the direct cause of many of these social deficits and, in the long run, of business inefficiencies as well. Without a healthy family and a community which sustains it, there is no healthy worker. Without a worker dedicated to the work ethic because over the years he has seen it provide for and protect his family and neighborhood, there can be no business efficiency, but only the encouragement of disillusioned workers exploiting managers and managers returning in kind.

What can we do to reverse this deterioration in family and community life and, subsequently, the deterioration of our business efficiency? What can we do to protect family stability and the work ethic which both expresses that stability and over the generations is nurtured by that stability?

Let me begin my concluding suggestions by reiterating my overall argument. I have claimed that free market ideology is based upon a view of motivation which sees humans as essentially selfish when operating in wider groups. I have suggested that this viewpoint, while obviously containing elements of truth, loses accuracy when overgeneralized, because it blocks from view our continuing sociality, our dependence upon others. Free market ideology encourages a short-sightedness that does not tally the costs of human connectedness—the quality of shared streets and neighborhoods.

Second is what I have called the theodicy of moral paradox—that social benefit comes unintended and, given the perversity of human acquisitiveness, *can come only unintended*. This paradox has foreclosed the conversation between economics and ethics, the first with its calculation of economic costs and the second with its concern for conscious moral choice.

This free market justification for the allocation of costs is inadequate because the market must begin to factor into its calculation of efficiencies not simply the freedom of capital transfers but also the many social costs which high velocity capital mobility entails. Economics can no longer avoid ethics. Nor can ethics any longer remain economically unsophisticated. A new dialogue must begin.

The key to that dialogue is the issue of how social costs can be brought to bear upon capital decisions. Lord Keynes, whom we have not quoted on our side before, nevertheless seemed to appreciate the need for an alliance between community and capital. Investment should be a domestic affair, he claimed; it should be "homespun."

> I sympathize with those who would minimize, rather than those who would maximize economic entanglements among nations.

> Ideas, knowledge, science, hospitality, travel—these are the things which should of their nature be international. But let goods be homespun wherever it is reasonably and conveniently possible, and above all, let finance be primarily national.[32]

How can we increase the accountability of capital to community? How can we enhance *the homespun*, the economy of the neighborhood, which is, at last, a nation's most important resource and defense?

Let me begin with the minimal needs and move, then, towards a more comprehensive program. There are a number of proposals for plant closing legislation which would help protect American workers from naked exposure to capital flight. The proposals are uniformly modest. They would require companies to provide a year's pre-notification of possible plant closings to let local leaders find alternatives to a shutdown and give communities time to prepare. The legislation would also require companies to provide severance pay to workers and to give the communities left behind readjustment funds to cover, for a while, lost tax revenues.

All this is not very much. At most it might slow a little the high velocity of capital mobility. It might cushion slightly the impact of plant closings upon workers, their families and their communities. But even then we would remain far behind other western industrialized nations in providing job security and a humane network of social services for workers thrown out of their jobs.

Longer range programs should be more systematic. For example, we presently require an "Environmental Impact Statement" before a company can build a dam or open a chemical plant. Why not legislate a *Social Costs Impact Statement*? This law should require a systematic assessment of the human costs of a proposed plant closing that would then have to be factored into the company's calculation of its capital efficiencies. This internalizing of the external costs proved effective in protecting the environment. Do blue-collar families and their communities deserve less protection? Are they any less an endangered national resource?

It is only under the discipline of such a law that U.S. corporations will begin to engage fully in worker retraining. Only when companies perceive that it is in fact too costly *to themselves* to leave workers behind in the transfer of capital to seek bigger profits elsewhere that serious effort will be given to updating the skills of a given workforce. Sweden, West Germany, and Japan already socially mandate a far higher degree of job security than we do. The result is predictable: a far more elaborate system of worker retraining and adjustment.

We are doing something very foolish in our society today. We are allowing socially unrestricted capital mobility to tear the financial and cultural roots out of working class neighborhoods. We are undermining the human meaning of work and reducing the work ethic to ineffectiveness. None of us can benefit from this kind of destruction.

Moreover, the proven importance of family and community means we need to use tax laws to encourage corporations to update factories where they are. The human uses of extended family and of stable neighborhoods mean *new jobs need to be directed to already established industrial areas*. At present, federal tax laws encourage just the opposite. They encourage the building of *new* plants to get the tax advantages of rapid depreciation. Corporate after-tax profits increase, but the community pays.

As a society we can't afford that unfairness any more. To barter away community in the name of productivity is irrational because the real "capital" of human productivity is family and neighborhood vitality. The operation of the market and of the tax system must be disciplined to the realism that the interests of profits and the interests of community must in the end be made to coincide.

I have argued that free market ideology is inadequate to our present situation in two ways. First, its individualism is more individual and less social than we humans are and must be. Second, its theodicy of moral paradox cuts the nerve between ethics and economics, which, precisely now, must support a lively trafficking if we are to stop the further erosion of our nation's communities. The costs of human connectedness must be tallied, not just those of individual greed. The market does not operate automatically to benign effect, not if one understands the social dimension of human life.

From neighborhoods we have learned how individual good is inextricably bound up with the common good. Good streets are shared streets. Decent blocks are common shelters and common productions of that shelter.

The free market, with its allocation of costs and its justification of sorrows, is no longer, if it ever was, morally or practically sufficient. Today, economics is and must be recognized to be as much ethical choice as technical expertise. We shall grasp this fact, or we shall watch morally unguided capital mobility destroy family and community stability, which is our nation's most important resource and the one impossible to replace.

Abraham Lincoln once said, "A house divided against itself cannot stand." Today, our nation is again becoming a divided house.

Notes

1. Quoted in Suzanne Keller, *Beyond the Ruling Class* (New York: Random House, 1963), p. 269.

2. Dun and Bradstreet figures quoted by Bluestone and Harrison, *Capital and Communities* (The Progressive Alliance; Washington, D.C., 1980), p. 59.

3. Draft Report of the McGill Commission, Nov., 1980.

4. "Theodicy" means, literally, the justification of the goodness and power of God in the face of evil and tragedy in life. I first developed some of the political implications of theodicy in my essay "Theodicy and Politics" (*Worldview*, April 1973). More recently, Jon Gunneman in an unpublished essay entitled "Market, Theodicy, and Ethics" has developed some of the economic implications. He forms very suggestively the idea of "moral paradox" in the underlying legitimacy system of free market ideology.

5. Adam Smith, *An Inquiry into the Nature and Causes of the Wealth of Nations* (1776), bk. V, ch. 2.

6. Quoted from E. F. Schumacher, *Small is Beautiful* (New York: Harper & Row, 1973), p. 24.

7. I am indebted on this point to Gunneman.

8. The best recent discussion of the idea of the "common good" is Robert Paul Wolff's *The Poverty of Liberalism*, especially his final chapter.

9. By "first" I mean only that point in the development of western civilization when public reflection came heavily under the influence of the Christian church, and its ideas about the nature and destiny of man.

10. St. Augustine, *City of God*, bk. XIII, ch. 27.

11. Max Weber, *The Protestant Ethic and the Spirit of Capitalism* (New York: Scribners, 1958), pp. 105–6.

12. Thomas Aquinas, *Of the Rule of Princes*, Vol. 1, p. i, trans. G. G. Phelan, extracted from D. Bigongari, ed., *The Political Writings of St. Thomas Aquinas*.

13. Thomas Aquinas, *Contra Impugnantes Dei Cultum et Religionem*, ch. 3, extracted from D. Bigongari, ed., *The Political Writings of St. Thomas Aquinas*.

14. I am certainly not arguing that Calvin intended any of this to happen. The relationship I am establishing here between the Augustinian/Calvinist view of human selfhood and the rise of free

market ideology is far more ironic. As does Max Weber, I see the linkup not as one intended by the theologians, who would certainly have been offended by the notion of unleashing human greed as an instrument to attain social benefit. But theological ideas and the ethos which they generate take on a life of their own, independent of their origins. They interact with other powerful developments in society, and in that combination produce consequences which were not, and in the nature of things could not have been, part of the original intention.

15. For a criticism of this approach, see Robert Paul Wolff, *Poverty of Liberalism*, and Theodore Lowi, *The End of Liberalism*. Also see the provocative reflections in Jon Gunnemann, "Market, Theodicy, and Ethics."

16. Gibson Winter's *Elements for a Social Ethic* has an excellent summary of this tradition. Also see, Berger and Luckman, *The Social Construction of Reality* and Clifford Geertz, *The Interpretation of Cultures*, esp. ch. 2 and 3.

17. Geertz, *Interpretation of Cultures*, p. 46.

18. I first developed some of these ideas in my *Worldview* article cited above.

19. The standard work in this area is Harvey Brenner's *Mental Illness and the Economy* (Cambridge: Harvard University Press, 1973).

20. Besides Brenner, see S. Lawrence, "Racial Discrimination—the Poor Relation," *Personnel Management* 7, no. 9 (1975): 22–26; and Banagale and McIntire, "Child Abuse and Neglect: A Study of Cases," *Nebraska Medical Journal* 60, no. 10 (1975): 393–96; also, B. Nichols, "The Abused Wife Problem," *Social Casework* 57, no. 1 (1976): 27–32; and Naomi Fiegelson, *A Child Is Being Beaten* (New York: Holt, Rinehart & Winston: 1975).

21. Besides Brenner, see Kraft and Babigian, "Suicide by Persons With and Without Psychiatric Contacts," *Archives of General Psychiatry* 33, no. 2 (1976): 209–15; also A. Pierce, "The Economic Cycle and the Social Suicide Rate," *American Sociological Review* 32 (1967): 457–62; also Henry and Short, *Suicide and Homicide* (Glencoe, Ill: Free Press, 1954).

22. Henry and Short, *Suicide and Homicide*; also Brenner, *Mental Illness and the Economy*.

23. Besides Brenner, see A. R. Bunn, "Ischemic Heart Disease Mortality and the Business Cycle in Australia," *American Journal of Public Health* (Aug. 1979), pp. 772–81.

24. Besides Brenner, see M. Droughton, "Relationship Between Economic Decline and Mental Hospital Admissions Continues to be Significant," *Psychological Reports* 36 (1975): 882.

25. See Richard Cohn, "The Effect of Employment Change on Self Attitudes," *Social Psychology* 41, no. 2 (1978): 81–93; and Catalano and Dailey, "Economic Predictors of Depressed Mood and Stressful Life Events in a Metropolitan Community," *Journal of Health and Social Behavior* 18 (Sept. 1977): 292–307.

26. The entire interview is published in *Christianity and Crisis* April 17, 1978.

27. See Walter Strange, "Job Loss: A Psychosocial Study of Worker Reactions to a Plant Closing in a Company Town in Southern Appalachia," Ph.D. dissertation (School of Industrial and Labor Relations, Cornell University).

28. Brenner, *Mental Illness*.

29. These figures are cited from the Bluestone essay in this book.

30. *Harvard Business Review* (Summer 1981).

31. Quoted from Douglas Bauer, "Why Big Business is Firing the Boss," *New York Times Magazine*, March 8, 1981.

32. "National Self-Suffering," *Yale Review* 22 (Summer 1933).

Index